About Island Press

Since 1984, the nonprofit organization Island Press has been stimulating, shaping, and communicating ideas that are essential for solving environmental problems worldwide. With more than 1,000 titles in print and some 30 new releases each year, we are the nation's leading publisher on environmental issues. We identify innovative thinkers and emerging trends in the environmental field. We work with world-renowned experts and authors to develop cross-disciplinary solutions to environmental challenges.

Island Press designs and executes educational campaigns in conjunction with our authors to communicate their critical messages in print, in person, and online using the latest technologies, innovative programs, and the media. Our goal is to reach targeted audiences—scientists, policymakers, environmental advocates, urban planners, the media, and concerned citizens—with information that can be used to create the framework for long-term ecological health and human well-being.

Island Press gratefully acknowledges major support of our work by The Agua Fund, The Andrew W. Mellon Foundation, The Bobolink Foundation, The Curtis and Edith Munson Foundation, Forrest C. and Frances H. Lattner Foundation, The JPB Foundation, The Kresge Foundation, The Oram Foundation, Inc., The Overbrook Foundation, The S.D. Bechtel, Jr. Foundation, The Summit Charitable Foundation, Inc., and many other generous supporters.

The opinions expressed in this book are those of the author(s) and do not necessarily reflect the views of our supporters.

MODERN POISONS

Modern Poisons

A BRIEF INTRODUCTION TO CONTEMPORARY TOXICOLOGY

Alan S. Kolok

Washington | Covelo | London

Library of Congress Control Number: 2016933395

Printed on recycled, acid-free paper

Manufactured in the United States of America

10 9 8 7 6 5 4 3 2 1

Keywords: Toxins, dose-response relationship, endocrine disruption, pesticides, chemical resistance, epigenetics, chemical regulation, persistent organic pollutants (POPs), Paracelsus, prions.

CONTENTS

PREFACE

Toxicology is interdisciplinary. Other disciplines, such as anatomy, can be studied more or less in isolation, without much intellectual investment from the other major scientific fields. Students can be educated on the arrangement of bone, muscle, and the internal organs, for example, with very little mention of the underlying chemistry of the bone, or the biomechanics involved in muscular activity. Toxicology, on the other hand, is the study of the adverse effects of noxious chemicals on living organisms, and therefore cannot be encapsulated solely within the fields of biology or chemistry, but rather lives within the intersection of the two disciplines.

Toxicology is also an applied science, being responsive to changes in the human environment and to societal needs. At its inception, toxicology was intricately associated with medicine. Physicians first developed the basic principles of toxicology over 500 years ago, for toxic insults were invariably personal and medically debilitating. The adages that "a chemical dose makes the poison" and "a chemical's nature is revealed through its structure" arose to help understand the mechanisms by which poisons were adversely affecting humans. During the early stages of its evolution as a discipline, toxicology borrowed and shared concepts with its chemically benign twin, pharmacology, the study of therapeutic effects of drugs on organisms. Indeed, the two fields are closely intertwined; the therapeutic nature of many drugs can turn toxic when the concentration of the drug's dose is too high, or when the exposure to the drug is maintained for too long.

Despite toxicology's historical tie with medicine, the changing nature of society's interaction with chemicals has changed the context in which toxicity occurs. Initially, most toxic exposures occurred on the individual level. Contact with poison ivy, the ingestion of an inedible mushroom, or a snakebite can all cause dire, and in some cases fatal, consequences; however, the chemical exposure is only directed at one person at a time. Even early historical efforts at metallurgy or manufacturing only affected those who worked at small and relatively isolated manufacturing sites. For much of human history, the exposure of humans to toxic chemicals was more personal than societal.

All of this changed with the industrial revolution. Large-scale soil, water, and air pollution began to occur as a result of heavy industries such as the production of iron and steel, as well as industrial mining and petroleum extraction. These industries distributed chemical risk beyond the individual level to that of the community. Epidemiology, the branch of medicine that deals with the incidence and distribution of disease conditions in defined human populations, joined forces with toxicology, and studies began to assess the relative risk that chemical exposures exacted upon human communities and populations.

The chemical revolution, a latter-day offspring of the industrial revolution, further changed the face of modern toxicology, as the diversity of chemicals released into the environment skyrocketed. Initially, chemical exposures were natural or were the products of relatively simple modifications of natural products, such as the enrichment of metals from ore or the burning of wood and fossil fuels. The advent of industrial organic synthesis, which occurred during the early twentieth century, ushered in a completely new suite of compounds into the toxicologist's world, and these new compounds were behaving in a very unusual manner. Rather than rapidly degrading in the environment, they persisted—in some cases, for decades. Furthermore, these compounds were also beginning to show up in the tissues of the most unlikely animals living in the most unlikely locations.

As chemical contamination increased in both scope and diversity, the very nature of the toxic response was also being turned on its head. Far from the acute and immediate damage of a bee sting or snakebite, evidence began to mount that chemicals were exacting long-term and subtle effects. Environmental pollutants, as well as drugs and personal care products, were acting as unintended cell signals, causing nefarious miscommunications among cells. Carcinogenicity, reproductive dysfunction, and alterations in fetal development were among the consequences. Ominously, these responses were occurring long after the chemical signal had vanished—at times, even affecting the children or grandchildren of exposed individuals. Like a thief in the night, the chemical intruder was long gone by the time the impacts were realized.

The triple threat of modern toxicology—the spread of chemicals globally, the spectacular increase in chemical contaminant diversity, and the complications presented by subtle yet enduring toxic responses—has greatly muddied the toxicological field. It is no wonder that there is considerable confusion regarding the nature of modern toxicology among laypersons, students, and scientists alike. It is not unusual for individuals to misunderstand the fate, transport, absorption, excretion, and biological action that govern toxic chemicals.

This newfound complexity in toxicology, driven by the events described above, is the impetus for this book. My purpose is make some sense out of the chaos, and to present the field in a framework that can be more clearly understood by the uninitiated reader. While some may view this book as being oversimplified, it is not meant to be the definitive text on toxic compounds, but rather to introduce the uninitiated to some of the subtleties and nuances within the field.

ACKNOWLEDGMENTS

This book is dedicated to three groups of people. First, my colleagues who helped with the early chapters of the book: Shannon Bartelt-Hunt, Sherry Cherek, Steven Ress, Christine Cutucache, Paul Davis, Eleanor Rogan, Philip Smith, Jeremy White, and Heiko Schoenfuss, thanks for the editorial help and comments. Second, my brother, sister, and mother, who kept asking me, in the nicest way possible, when the book would be finished. Well, family, after years in the making, it finally is. And finally my wife and son, Wendy and Jared, thanks for all the patience. More than anyone else, they know of what I speak.

Chapter 1

The Dose Makes the Poison

All things are poison, and nothing is without poison;
only the dose permits something not to be poisonous.
— Paracelsus

When I was in elementary school, conversations on the playground often took a fatalistic turn. Perhaps it was just an echo of the Cold War era, but I can recall chatting with my school chums about chemical compounds and death. We'd exclaim, "If you breathe too hard, it could kill you," or "You could drink so much water that you would die!"

Today, the concern is less about the lethal quantity of these relatively benign substances, but rather about pollutants in our food, water, and air. Yet despite our lack of sophistication, my young friends and I weren't actually so far off the mark. We didn't realize it at the time, but we were channeling a sixteenth-century physician, Paracelsus. Considered the father of toxicology, Paracelsus is credited with the first and most important tenet of the field, the idea that the dose makes the poison: "All things are poison and nothing is without poison; only the dose makes a thing not a poison." In other words, seemingly benign substances like water as well as obviously dangerous ones like arsenic can be deadly when administered in excess.

Paracelsus's groundbreaking idea centers on the dose–response relationship: the fact that in most cases the greater the dose, the greater the adverse, or toxic, response. While humble in its simplicity, the concept provides a thematic platform upon which modern regulatory toxicology is based. Furthermore, the relationship is actually more interesting than it would first appear, as both dose and response are surprisingly nuanced.

When a chemical, toxic or benign, contacts a biological organism, the contact is known as an *exposure*. The exposure dose is the quantitative amount of a chemical that a person (wittingly or unwittingly) is exposed to, and this quantity can be either directly or indirectly measured. For common chemicals that are deliberately administered, such as pharmaceuticals, the route of administration is direct, and generally occurs via oral consumption or injection. For exposures of this type, the dose is generally given in terms of the mass (in grams, g, or milligrams, mg) of the chemical being administered. For example, a regular-strength aspirin pill, one of most commonly consumed pharmaceuticals, contains 325 mg of the active ingredient, acetylsalicylic acid. The tablet also contains a number of other inert chemicals, but the *dose* refers to the amount of the active ingredient. For injections, the dosage is expressed in the same way. An epinephrine auto-injector, for example, widely self-administered by individuals with food allergies, will administer a dose of 0.3 mg of epinephrine to the individual despite the fact that the injected solution contains other chemical compounds.

In the examples given above, the exposure route is direct and easily quantifiable, but what if, on the other hand, the exposure is indirect? Indirect exposures would include the exposure that results when a fish ventilates contaminated water across its gills, or a person inhales secondhand smoke into their lungs. In these cases, the quantitative dose of the chemical exposure is much less certain, and much more difficult to measure. Rather than determining the exposure dose, it is far easier to quantify the concentration of the chemical in the "environment" (the water that the fish is ventilating, or the air that the animal or person is breathing). Furthermore, since the amount of the compound that the animal ventilates or inhales is not known exactly, the exposure cannot be quantified in terms of mass, but rather is quantified in terms of its concentration (the amount of chemical found in a specific volume of air or water) in the local environment.

Regardless of the direct or indirect source of the exposure, the response of an animal to a chemical exposure is also generally expressed in one of two broad categories, either discrete or continuous. Organism death is the ultimate discrete response, in that animals can only be found in one of two states, dead or alive. While perhaps somewhat gruesome, death provides a very valuable (and oft-times used) endpoint for toxicological studies. In contrast, variable responses to an exposure can also occur. For example, the impairment of cognition due to alcohol consumption is a classic example of a continuous variable. The response to alcohol is not all-or-none, but rather increases in its impact as the administered dose increases. This is also true for other types of toxicological impairment, such as changes in genetic expression or alterations in the activity of proteins.

Interestingly, the way that an exposure dose is expressed, whether indirect or direct, and the way that the response is measured, whether discrete or continuous, do not affect the overall shape of the dose–response relationship. In the majority of cases, the shape of the dose–response curve remains sacrosanct regardless how the dose and response data are represented within it.

Quantifying the Dose–Response Relationship

The dose–response relationship is a very powerful tool, frequently used by regulatory agencies. A common approach used to test new chemicals, or chemicals used in novel ways, begins with the generation of dose–response relationships. Generally, the first battery of toxicity testing evaluates the capacity of a chemical to produce the discrete endpoint, death, which is exacted upon a population of experimental laboratory animals, such as mice.

A dose–response curve does not really focus upon death, but rather mortality. *Death* is the response of an individual organism, and clearly each individual can be in only one of two states: dead or alive. In contrast to death, *mortality* is the response of a population of individuals. The *mortality rate* describes the proportion of a population that dies in response to a calamitous exposure to toxic chemicals. To graphically illustrate the mortality of a group of animals that are exposed to the same dose of a toxic compound, we use the discrete dose–response curve. At one extreme of the toxicology curve, animals exposed to low doses survive (mortality rate is zero), whereas at the other extreme all of the animals exposed to higher doses of a chemical die (mortality rate is 100 percent).

In between total survivorship and total mortality, the dose–response relationship gets more interesting. In the vast majority of cases, the relationship between the two is a characteristic "S" or sigmoidal shape. At low chemical doses, an incremental increase in the concentration of the toxic substance does not lead to a very large increase in mortality. At intermediate doses, the impact of the compound on mortality increases dramatically, while at the highest doses, the increase in mortality from one dose to the next higher dose is again minimal.

An important point to help clarify the relationship is the *inflection point*. On the lower half of the curve, increases in dose lead not only to a greater number of animals dying, but also to an increase in the rate at which mortality increases from one dose to the next. In other words, the slope of the line from one concentration of a chemical to the next continues to increase until it reaches a maximum slope at the inflection point. Further increases in the dose of the compound continue to elicit a greater biological response, but the rate at which the response increases is now declining with each successive increase in dose administered. The inflection point always occurs at the midpoint of the curve, the point at which 50 percent mortality would occur in a lethal toxicity test, and, as will become apparent in a later section of the chapter, the inflection point has taken on considerable importance with respect to toxicological testing.

The transition from experimentally derived data to a useful dose–response relationship (one that allows points of interest along the curve to be quantified) is more difficult to come by than it may first appear. Filling in the gaps between a few, relatively scarce data points (experimentally collected) to a complete curve, requires the use of a mathematical equation that characterizes the relationship. Once that equation has been defined, it can then be used to identify any point along the curve, not just points where data has been collected.

Pragmatically, there are important experimental design issues that have to be resolved when elucidating the dose–response relationship for a chemical compound. For example, if a toxic compound is novel and has never been tested previously, then the researcher is flying blind and will need to generate a dose–response relationship that includes a wide range of chemical concentrations. Very often the range is so large that the x-axis of the dose–response relationship is not represented arithmetically (that is, 1, 2, 3, and so on) but rather is arranged geometrically (that is, 1, 10, 100, etc.). In this case, it is highly likely that the experiment will include one or more groups of animals that are exposed to chemical concentrations that generate no mortality, and one or more doses that cause total mortality. Importantly, these doses do not help to quantify the dose–response curve. After excluding these points from analysis, the number of data points remaining to assess the sigmoidal curve may become disturbingly small, thereby reducing the scientific confidence that the researcher may have in the results.

Fortunately, there are mathematical methods by which some of these difficulties can be circumvented. *Probit analysis* allows for mathematical gyrations to occur so that a sigmoidal relationship can be straightened into a line. As students of Euclidian geometry can testify, the shortest distance between any two points is a straight line, and conversely, any line can be described by only two points. As such, the entire dose–response curve can be accurately estimated using probit analysis when as few as two of the chemical doses provide data that lie somewhere between zero and total mortality. Furthermore, once the relationship is described by a linear equation (y = slope*x + y-intercept), any point on the line can be readily quantified by plugging a few numbers into the simple linear equation.

Dancing along the Dose–Response Relationship

The beauty of the linear dose–response relationship is that it provides a wealth of preliminary information regarding the interaction between the animal and the chemical. For example, the slope of the line provides information regarding the efficacy, or the capacity to produce a biological effect, of the toxic chemical. As the slope increases, the efficacy of the chemical compound also increases. Furthermore, if the efficacies of two compounds are similar, then dose–response relationships can yield a number of useful points that provide a shorthand, a single number, by which the toxicity of the different chemicals can be compared.

Now recall the inflection point that was discussed previously. The inflection point, known as the LD_{50}, is the chemical concentration at which 50 percent of the animals die due to the exposure. If animals are exposed to an environment (a noxious gas in the atmosphere for animals that breathe air, or a toxic compound in the water where fish live), the inflection point can still be evaluated, although it is given the moniker LC_{50} (the chemical concentration in the organism's environment at which 50 percent of the organisms have died). These inflection points provide a handy numeric index that can be used to compare the toxicity of different compounds.

The second point that can be gleaned from a dose–response relationship is the threshold concentration. The *threshold concentration* is the concentration at which the probability of an adverse impact (for example, one adverse case per million individuals) is low enough to be deemed acceptable. Interestingly, while the threshold dose can easily be located on the dose–response line once the acceptable rates of adverse impacts are agreed upon, the acceptable probability of adverse effect is socially or politically determined, as opposed to scientifically determined. This topic will be discussed in greater detail in chapter 14.

A third important point regarding the threshold dose is that it is a mathematical rather than an empirical construct. In other words, a threshold dose is not limited by the choices of the scientist conducting the test. If a scientist, for example, has injected rats with a chemical at five concentrations (0.01, 0.1, 1, 10, and 100 milligrams per kilogram), the threshold dose is not limited to those concentrations. The line derived from a probit dose–response relationship does not just describe the relationship for a few points on the line, but rather describes the relationship for all of the points on the line. This is not the case for two other commonly used endpoints, known as NOEC and LOEC.These endpoints have also been used to infer chemical safety, despite the fact that both metrics have fallen into considerable disfavor. The *NOEC* (no-observable-effect concentration) represents the greatest measured chemical concentration on the curve that does not yield a positive effect, whereas the *LOEC* (lowest-observable-effect concentration) represents the lowest measured concentration of the chemical that yields a biologically adverse effect. Importantly, these metrics are inherently biased. While the threshold dose is mathematically derived, using all of the points on the dose–response curve, NOEC and LOEC values only correspond to the points on the curve where the exposures were empirically conducted. As such, the number of exposure doses selected by the experimenter limits the total number of possible values for these points. For example, the scientist that injected rats with a chemical at five concentrations (0.01, 0.1, 1, 10, and 100 milligrams per kilogram) can only determine NOECs or LOECs at one of the five concentrations at which the experiment was actually conducted. Effectively, the results, rather than being mathematically derived and unlimited, are derived by the whim of the investigator and are extremely limited.

When a federal or state agency develops chemical safety standards, these are almost always lower than the threshold values generated from the dose–reponse relationship. The reason for this is purely pragmatic, as the results from toxicity testing are generally derived from rodent models (rats and mice) and applied to humans. Rodents can be either more or less sensitive to the chemical than humans; therefore, a safety factor is often applied in order to decrease the maximum contaminant level by an order of magnitude—in other words, a tenfold reduction in the contaminant-level goal. Furthermore, while one safety factor can be applied to safeguard against species differences, a second factor can be applied to take into account the enhanced risk of the chemical to sensitive subpopulations (infants, children, the elderly, and those with compromised immune systems).

Exceptions

But what if the sigmoidal relationship is not valid? For example, consider vitamin A. Vitamin A is actually a suite of compounds, including retinol, retinal, and a number of similarly structured carotenoids. An insufficient intake of dietary vitamin A leads to a deficiency that can cause impaired vision, particularly during low light levels. Yet vitamin A is a fat-soluble compound that cannot be excreted as readily from the body as water-soluble vitamins, such as vitamin C. Therefore, if one consumes too much vitamin A, there is a risk of toxicity. Chronically high levels of vitamin A are toxic, particularly with respect to fetal development during organogenesis, the time of development when the primary body organs are developing. In the case of vitamin A, the dose–response curve is not a sigmoidal curve that consistently slopes forward, but rather a bowl shape with adverse impacts occurring on the lower end, where vitamin A deficiencies occur, and on the upper end of the distribution, where overt toxicity occurs. In this case, the dose makes the poison in two different ways: at higher concentrations of vitamin A, the toxic effect prevails and the lower the level the better. At low concentrations of vitamin A, the compound acts as a micro nutrient, deficiency governs the impact, and the more vitamin A the better.

While deficiency is not really a toxic impact, there are examples of other chemicals where adverse impacts occur at both lower and higher exposure concentrations. The compound 17β-estradiol is a perfect example. At high levels, 17beta-estradiol increases the risk of carcinogenesis, and can be overtly toxic. However, as the dose decreases, the impact will attenuate according to a classic dose–response sigmoidal relationship. Doses lower than the threshold will not increase cancer risk. However, at doses much lower than the threshold, this sex steroid will also act as a cell signal that helps to govern vertebrate fetal development, among other functions. While estradiol is essential to the development of both males and females, unusually elevated levels of the compound at the wrong times can lead to toxicity, including inappropriate female pattern development in males. This is the genesis of unusual reproductive morphologies in some male animals, such as ovo-testes in which ovarian follicles develop within the testicular tissue of males. (The relationship between chemicals and fetal development will be revisited in later chapters of this book.)

Our view of toxic chemicals is governed by the edict of Paracelsus, a sixteenth-century physician. For many toxins, perhaps the vast majority, the dose does make the poison. This first law of toxicology has driven a great deal of research and safety regulation. In the world of modern poisons, many of the interactions between the molecule and the organism are defined by this simple yet elegant relationship.

Chapter 2

The Nature of a Chemical

Some things just aren't meant to go together.
Things like oil and water.
Orange juice and toothpaste.
— Jim Butcher

The second rule of toxicology helps to explain why some chemicals are easily excreted from the body, while others are not. It also goes to the heart of why different toxic agents affect us in different ways. The second rule was first posed by Ambroise Paré, a sixteenth-century French surgeon, who realized that "Poison . . . kills by a certain specific antipathy contrary to our nature." In other words, a chemical's particular action depends on its inherent chemical nature.

This is a seemingly simple concept, but what exactly is meant by a chemical's "inherent nature"? We now know that chemical compounds generate biologic effects based upon the structure of individual molecules. Toxicity occurs, by definition, at the molecular level, as individual toxic molecules bind to individual biological target sites to generate effects. These biological target sites can be general, such as the thin layer of fat, the phospholipids, that comprise the membrane that encircles every cell, or they can be very specific, such as a neurotransmitter receptor site being deactivated by its irreversible binding to one of a very specifically configured chemical pesticide. Therefore, the inherent nature of a toxic chemical resides within its molecular structure, for it is the molecular structure that predicts the chemical's activity.

For many toxic compounds, particularly the "specialists" that bind to very specific cell receptor sites, it would stand to reason that the three-dimensional structure of the toxic compound is responsible for its cellular impact. If the toxic compound is a key, then the receptor molecule is a lock, and for many toxic chemicals, differences in the chemical structure of the molecule (analogous to slight differences in the cutouts on the blade of a key) can lead to alterations in its toxicity. Furthermore, if the toxic mechanism by which a compound elicits a biological effect is understood, it stands to reason that molecules that share a similar chemical structure may cause similar effects when organisms are exposed to them. The structure–activity relationship, the correlation between the structure of a chemical compound and its biological activity, is the major corollary that can be derived from Paré's writings.

Relative to the specialist toxic compounds, the structure–activity relationship explains why the various members of some small families of chemicals all produce the same impact. But what if we were to take a more comprehensive view of chemical compounds? If all known chemicals could be drawn on a blackboard (it would have to be a pretty large blackboard!), they could be organized into a large number of small family groups:the sugars on one part of the board, the elemental metals on another, the dioxins on yet another region of the board, and so on. If we were to look at the blackboard as a whole, what segregation criteria could be used to begin to split the board up into the very largest nations of chemicals?

There are two very useful and elementary ways that the board could initially be divided. The first would be to divide the organic molecules from the inorganic. This would simply segregate those compounds that contain carbon from those that do not. In general, carbon-containing compounds are those that can be acted upon, in other words modified chemically, by living organisms, and they are generally considered fundamental to life. All of the molecules that are essential to life—the sugars, fats, and proteins, the DNA and RNA, the lipid membrane, and on and on—are all organic as they all contain carbon.

The second would be based upon the relative solubility of the compound; in other words, whether the molecular is water- or lipid-soluble. Unlike the organic/inorganic dichotomy, which is strictly either/or, the relative solubility of compounds ranges along a continuum from highly lipid- or fat-soluble to highly water-soluble. Nevertheless, a distinction between water-soluble and lipid-soluble compounds is instructive, as it explains so much.

The distinction between water and lipid solubility is as easy to visualize as is the process of making salad dressing. In a traditional Italian salad dressing, water, olive oil, and spices are combined in a decanter and then shaken vigorously. The resulting solution is actually an emulsification—that is, a mixture of two solutions that cannot be permanently blended together. Over time, the emulsified liquids will segregate and the salad dressing will separate so that a layer of olive oil will be found riding on top of the underlying water. Now think about the spices that were used in the salad dressing: on a molecular level, some of them (table salt, sugar, etc.) will primarily dissolve (or become soluble) in the water rather than the oil. Others, such as vanilla, mint extract, wintergreen extract (okay, it's a non-traditional salad dressing) will dissolve primarily into the olive oil, and virtually none of these lipophilic spices will solubilize, or dissolve into the water.

Now think about adding any chemical compound listed on our imaginary blackboard into the oil/water concoction. If we shake the mixture vigorously, then allow it to resegregate, the compound will partition into one of two compartments—the oil or the water—or it will not partition at all. That is, it will dissolve in the water, dissolve into the oil, or not dissolve at all and be found sitting, in crystal form, on the bottom on the decanter.

Based upon the two distinctions among chemicals we have made, virtually all of the known chemicals can be placed into one of five categories: insoluble compounds, lipid-soluble inorganic compounds, lipid-soluble organic compounds, water-soluble inorganic compounds, and water-soluble organic compounds. The insoluble compounds, both inorganic and organic, are held together by strong chemical bonds that cannot easily be broken apart; therefore, they do not contribute to the chemical composition within the lipid or water. Considering the fact more closely, if a compound is not water- or lipid-soluble then it cannot be absorbed into the body, and without absorption there is no toxicity. As far as this chapter, and indeed this book, is concerned, insoluble compounds are just not very interesting toxicologically.

This reduces the classification of a toxic compound to one of four categories: lipid-soluble and water-soluble inorganic compounds, and lipid-soluble and water-soluble organic compounds. The list becomes even smaller, as lipid-soluble inorganics are not a chemical class that is of much interest toxicologically.* In many ways, the dichotomy between lipid and water solubility is as important as the split between inorganic and organic. To understand that, it is necessary to take a closer look at the point where chemicals literally bump up against biology: the cell membrane.

The Cell Membrane

Before contemplating the absorption of compounds from the environment into the body, take a moment and consider the cell. When teaching general biology for non-majors one semester, I asked my students to pull out a sheet of paper and, in the time provided, draw

* This distinction is problematic for inorganic ions such as mercury. Mercury is inorganic and water-soluble as an ion, but it can be complex with organic compounds that are lipid-soluble.

a cell. It was a sneaky question, as I only gave them a second or two to begin the illustration before yelling out, "Time!" In that very brief time period, most students drew the same thing, a closed circle. In essence, what they drew was the cell membrane. While the nucleus, mitochondria, golgi apparatus, and so forth may all be vitally important to the life of a cell, it is the cell membrane that separates the inside from outside, and it is this humble cell structure that is first drawn by virtually everyone who is given the challenge of drawing a cell in three seconds or less.

The cell membrane is composed of a bilayer of phospholipids. The phospholipids are arranged in such a manner that, in the outer layer, the phospholipid molecules' polar heads point outward from the cell and their nonpolar tails point into the cytoplasm. After a small gap, the inner layer of phospholipid molecules in the bilayer lies with their tails closely associated with those of their outer partners, and their heads abutting the cytoplasm of the inner reaches of the cell. As such the lipid bilayer is composed of polar heads inside and out, with nonpolar tails in the middle of the membrane.

The polar heads are critical to this arrangement. Within each of the polar heads lies a positive and a negative region. As such the outer membrane surface is composed of a large number of heads that are both positively and negatively charged. Since like charges repel and opposite charges attract, the lipid bilayer represents an impenetrable barrier to diffusion for any charged inorganic or organic ion.

But this relative impermeability creates a problem. The inner cytoplasm cannot wall itself off from water-soluble compounds entirely. Many essential ions—micronutrients such as sodium, calcium, and chloride, and organics such as glucose or other sugars—are polar molecules and thus cannot readily migrate through the cell's lipid bilayer. Then how do they get in? They do so with the help of proteins. A wide variety of proteins stud the lipid bilayer, like so many rhinestones on an ornamental belt. The proteins act as pores or carriers that can ferry polar molecules across the cell membrane. In fact, if one surveys the degree to which biologic membranes are studded with proteins, one finds that the most-active structures (mitochondria) have the most well-endowed membranes, fully studded with proteins, whereas the least metabolically active structures (e.g., the membrane of insulation that wraps around portions of nerve cells), are particularly devoid of this protein-based studding.

For a toxic response to occur, the chemical has to reach its target. Sometimes the target is a receptor, other times it is a protein or the nuclear DNA, but generally speaking the target is either within the cell, embedded within the cell membrane, or the cell membrane (lipid bilayer) itself. As such, for many toxic chemicals the route to biological activity has to transcend the cell membrane, and that is where the solubility of the chemical reenters the picture. Water-soluble compounds (organic and inorganic alike) cannot readily cross the lipid bilayer that makes up the cell membrane unless they have help in the form of a protein gate, pore, or carrier. As such, water-soluble compounds are controllable and many—for example, the inorganic ions such as sodium, chloride, potassium, and calcium—are maintained in strict concentrations within the cell.

Interestingly, while this system has evolved so that these ions can be closely regulated, the system is not so finely tuned to assure that mistakes do not occur. Ion channels or gates allow for the precise regulation of inorganic or organic ions but also inadvertently allow for the inappropriate uptake of toxic ions from the blood into the cell. Transporters for the micronutrients copper and zinc cannot differentiate between these essential metal ions and more insidious metals, such as cadmium, silver, and mercury.

While toxic water-soluble chemicals engage in a game of mistaken identity, being allowed into the cell via channels whose function is to allow the transport of necessary chemicals, toxic lipid-soluble organic molecules are engaged in a very different process. These compounds do not see the lipid barrier as a boundary, and as such can move around within an animal without constraint. In essence, these molecules are mavericks, capable of coming and going within the organism, in an unregulated sense. Considering that one definition of a living organism is an entity that controls its internal composition; the lipid-soluble molecules make an end run around the order and organization that is inherit within living cells.

Since solubility is so important to the absorption and ultimate fate of molecules within the body, we need a way to quantify it. This brings us back to the salad dressing analogy. If a mystery compound is added to a slurry (emulsion) of water and oil, and the slurry is allowed to settle, with the oil rising to the top, all one needs to do to determine the solubility is to measure the concentration of the compound in the oil and water layers. When done experimentally, the oil used when quantifying the level of solubility is octanol and the resulting numeric determination is referred to as the octanol:water partition coefficient (K_{ow}).

If we think about different molecules that range from highly water-soluble (table salt) to highly lipid-soluble (cholesterol), how different are their K_{ow}s? It turns out that they are very different indeed. It would not be unusual for a water-soluble compound to be a million times more soluble in water than oil, nor would it be unusual for a lipophilic compound to be a million times more soluble in oil than water. In fact, these numbers are so large that K_{ow}s are usually represented as logarithmic functions (the K_{ow} of a lipid soluble compound may be 1,000,000 or 10^6 or $\log K_{ow} = 6$), and indeed the difference between the solubility of lipid- and water-soluble compounds may exceed 10^{12}, such that the solubility of a lipid-soluble molecule into a lipid bilayer can be more than one trillion times that of a water-soluble compound relative to the same bilayer.

Now we can reexamine the lipid bilayer as a chemical barrier, looking at it from the perspective of K_{ow}. A chemical that is water-soluble is, by definition, not lipid-soluble and as such will not be able to dissolve into the membrane. As such, its rate of diffusion across the barrier will be minimal. Without the aid of carrier proteins, the uptake of the chemical will be minimal, as will its toxicity. In contrast, a compound that is soluble in oil or fat is easily absorbed across the lipid bilayer, and as such, will have a greater potential for eliciting a toxic action. With a high log, K_{ow} will readily solubilize into the lipid bilayer and its rates of diffusion into the cells will be much higher.

So there it is, perhaps the fundamental dichotomy within toxicology: the issue of water-solubility. The behavior of virtually every toxic compound is directed by the fundamental dichotomy of solubility: water versus lipid. Compound solubility influences fundamental processes such as absorption from the environment, delivery in the blood, diffusion into the target tissues, excretion from the target tissue, metabolism, sequestration, and whole-body elimination. These functions are not as strongly regulated by molecule size or shape as they are by the relative lipid/water solubility. Contaminant solubility, the road to toxicology, begins here.

Chapter 3

The Human Animal

All animals are equal, but some animals are more equal than others.
— George Orwell, *Animal Farm*

Like a scientist in a white lab coat, a laboratory mouse instantly conjures up images of medical research. Yet despite the mouse's iconic position in the scientific community, it may soon be replaced by the zebrafish as the laboratory animal of choice. Why might this cold-blooded fish prove more valuable than the furry little warm-blooded icon, and how can this fish species, so different from our own, play a role in medical or toxicological research?

The answer begins with the realization that humans are animals. And all known animals, human and nonhuman alike, share many similarities. First of all, we are all heterotrophs (consumers of organic material) and are all capable of locomotor activity (movement). Even our basic structure is the same: we are all multicellular and package our genetic material in a nucleus, a specific and well-defined region within our cells. On the molecular or biochemical level, the processes of respiration, digestion, excretion, and metabolism (both catabolism, or the breakdown of foodstuffs into elemental building blocks, and anabolism, or the production of tissue from these basic building blocks) are remarkably consistent among the various animal species. The considerable overlap among animals provides the basis for animal testing: if we know how a particular toxic compound affects an animal species such as a laboratory mouse or a zebrafish, that knowledge may also give us clues regarding how that chemical will affect humans.

Determining what species are most useful for medical research depends upon a number of factors, including society's attitudes toward different kinds of animals. Specifically, we tend to regard the vertebrates with more empathy than the animals without backbones. For example, until very recently the Institutional Animal Care and Use Committee (IACUC), the administrative body that governs the use of animals, focused primarily on vertebrates, and more specifically, on mammals. Mammals are an obvious focal point for IACUC because rodents are so frequently used in biomedical research. But there is more to the rationale then mere familiarity or sheer numbers would suggest. Part of IACUC's charge is to alleviate unnecessary pain and suffering in laboratory animals. The more evolutionarily distant from us animals are (that is, the longer back in time one must go to find a shared or common ancestor), the more difficult it is for us to understand what, if anything, they feel.

The fact that invertebrates (snails, worms, insects, clams, and so on) are so morphologically different from us makes it relatively easy to draw a line in the sand and separate the "true animals," the vertebrates, from the "pseudo-animals," the invertebrates. In fact, our language mirrors the lack of empathy, and indeed revulsion, that many hold for the bulk of invertebrate animals. Our disdain is evident in common insults, such as "He's spineless" or "She needs to get a backbone," comments that separate the invertebrates from us and our vertebrate brethren.

Regardless of our human biases, the similarities between humans and animals do not end at the dividing line between the vertebrates and the invertebrates. For while it may be difficult to determine whether the pain and suffering of an animal "without a face" is consistent with that experienced by a mammal or a vertebrate, the similarities of organisms—functionally, metabolically, and molecularly—cannot be overlooked. These similarities have led, in comparative physiology, to the establishment of Krogh's principle, proposed by August Krogh, winner of the 1920 Nobel Prize in Physiology. Krogh's principle states that "for such a large number of (physiological) problems there will be some animal of choice, or a few such animals, on which it can be most conveniently studied." Krogh's principle applies to all animals, not just the ones with backbones.

Krogh Meets *Loligo*

Examples of the utility of Krogh's principle abound. In some systems, invertebrates may be simpler then vertebrates, but they may also have unique or larger structures, or they may have structures that possess properties not readily exhibited by vertebrates. A classic example of this phenomenon are the giant axons in the squid *Loligo*.

In all animals, electrical information travels from the spinal cord to the muscles through nerve bundles. These bundles of nerves are very much like coaxial cables in that there are individual cell processes, or axons, that run from nerve cell bodies (neurons) in the spinal cord to muscle fibers in the limbs. These axons can be long (over a meter) but are very small in diameter, resembling a wire or physical conduit from the neuron that reaches a target region some distance away. Considering that axons extend from a cell body in the nucleus that is only 10–25 micrometers (10^{-6} meters) in diameter, these cell extensions are very fine threads of electrical conduit.

Speed of locomotion depends on, in part, the speed at which the nerve signals travel along axons. To increase this speed, vertebrates have insulated the wires with a metabolically inactive membrane. Invertebrates, by contrast, have increased speed by increasing the diameter of the axons themselves. Both of these advancements reduce resistance to the flow of electrical charge. In the giant squid, *Loligo*, this increase in diameter (found within the coaxial cable known as the stellate nerve) is taken to an extreme, as a single axon is between 300–800 micrometers in diameter. In other words, the axon, the membranous extension of a single nerve cell, or neuron, has increased in diameter to the point where it is at least 12 times the diameter of the average neuron within the mammalian spinal cord!

This is important to physiologists, as this giant axon is comparatively easy to manipulate. Experiments with this unique and enormous structure allowed early twentieth-century physiologists to determine the mechanisms underlying how electrical signals are transmitted by nerve cells. Krogh's principle is aptly applied to this system, as the giant squid has proven to be the organism of choice for studying neuronal transmission of information in all of the other animals, vertebrate and nonvertebrate alike. Of course, there are differences between the axons of squid and those of mammals, including the innovations involved in increased transmission velocity. Nevertheless, the fundamental principles for how these axons function transcend the artificial dividing line between vertebrates and invertebrates.

The Read-Across Hypothesis

The similarity between humans and other animals is useful not only for developing model organisms for medical research, but also when attempting to determine the adverse impacts of chemicals released into the environment. As will be discussed in a later chapter, human pharmaceuticals and personal care products are entering the environment, very often in complex mixtures. Given that many of these compounds have not been considered as environmental toxicants, and that the concentrations of the compounds in the environment are very low, the environmental risk is difficult to ascertain.

The conceptual approach known as the read-across hypothesis can be useful when considering the environmental risks posed by these compounds. Pharmaceuticals and personal care products may have environmental effects on non-target tissues that are consistent with the effects on humans, providing that the molecular target for the compounds is conserved. If this is the case, it can be assumed that the compound in the blood of the non-target species will develop a phamacological response at concentrations lower than those at which it develops a toxic response. Furthermore, if this is true, data derived during the the drug development process will be informative when considering the adverse toxicological impacts on non-target animals in the environment. The important point is that animals may respond to pharmaceuticals in a manner similar to humans, providing that they share the same intracellular molecular targets.

Genes and Genetic Variation

The reason that animals quite different from each other in appearance nevertheless share complicated functional structures, such as the nerve axon, is because they share a common ancestry. In animals, the conduction of locomotor signals in animals from brain to muscle evolved only once, and it has remained basically unchanged, or evolutionarily conserved, ever since. Certainly, there have been many modifications to the system since vertebrates and invertebrates split from each other (for example, the innovations necessary to increase velocity, as discussed previously). However, the fundamental molecular mechanisms by which the neurons work have, to a large degree, been conserved ever since the first animal neural network was organized.

The same principle holds for many metabolic enzymes, the proteins involved in the acquisition of cellular energy in the form of adenosine triphosphate, or ATP. In the early phases of sugar metabolism leading to energy production, the activation of a key protein (A-kinase) is regulated by a protein subunit that deactivates the enzyme when attached. Attachment of the subunit to A-kinase is governed by an internal cell-signal molecule, cyclic AMP (or cAMP), and when this binds with the regulatory unit it is released from A-kinase, thereby allowing the enzyme to function. This metabolic level of control has been so well conserved throughout evolutionary time that the regulatory subunits from one animal will bind with the A-kinase from a wide variety of other, unrelated, animals. The metabolism of sugars, glycolysis, is a fundamental pathway in all animals. As such, the proteins involved in it are highly conserved. Humans, being animals, share in this conservation, and the regulatory subunit from other animal species will readily bind with the A-kinase of humans.

These examples illustrate that, on a molecular level, humans are indeed animals. This is critically important to toxicology, as a chemical compound's impact on a target tissue is likely to be consistent across a wide variety of unrelated animals, including humans.

Natural Selection and Differential Susceptibility

If conservation of molecular structure and function is a tenet of biology, so too is the idea of variability among individuals. Individuals of the same nonhuman species, which may appear indistinguishable to us, can differ dramatically from each other with respect to their genetics, biochemical or physiological function, morphology, and behavior. The driver of these changes is natural selection, and the basic principles of natural selection are that (1) animals produce more offspring that can survive on the available resources, (2) an animal's progeny will vary in their morphological and biochemical attributes, and (3) the variability among the progeny will determine which of them is able to survive to successfully reproduce and thereby contribute to the next generation. Therefore, while many essential genes and proteins are conserved over time, this conservation occurs in the face of a continual reshuffling of the genetic deck in the progeny of each successive generation.

Individual variability among animals from the same species is important to toxicology in a number of different ways. First of all, it is responsible, to a large extent, for the shape of the dose–response curve. The genetic and morphological variation in the individuals that are used to elucidate the dose–response curve is responsible for their capacity to respond to a toxic exposure. For short-lived animals that tend to be numerous in the environment, such as insects, individual variation in chemical susceptibility can lead to the development of pesticide-resistant populations. Pesticide resistance can arise within an insect population quickly, as a pesticide will kill the susceptible insects, leaving the resistant ones behind to reproduce themselves. Therefore, the dose–response curve is not static over time, but changes based upon the biological response of animal populations to the chemical exposure.

The pervasiveness of variation in biological systems also influences the utility of laboratory model organisms in disease diagnostics, for the longer two species have been separated from each other in evolutionary time, the more different their molecular makeup. This becomes problematic when animals are used in laboratory medical testing. For example, the two most common laboratory animals, the lab mouse and the lab rat, have not shared a common ancestor for approximately 23 million years, and rodents in general have not shared a common ancestor with humans for approximately 100 million years. That separation has provided ample time for evolution to drive the animals apart, not just in their body form, or morphology, but also in their capacity to respond to chemicals. For example, when the chemical carcinogenicity of 392 chemicals were tested in both mice and rats, 76 percent of the rat carcinogens were also carcinogenic in mice, while 70 percent of mouse carcinogens proved to be carcinogenic in rats.

Somewhat paradoxically, while fundamental physiological and molecular processes allow us to use animal models in toxicity testing, selection pressures may have acted on both humans and the model organism to such a degree that the model's toxicological response is profoundly different from a human's. Given the greater evolutionary distance between humans and rodents than that between rats and mice, it is a safe bet that the chemicals found to be carcinogenic in humans and rodents alike are not much greater than the 70 percent mentioned above. This is not to imply that rodent models have no value; however, it does emphasize that point that both molecular conservation and natural selection are simultaneously working on all animals, including humans, all of the time.

Humans are animals, and our shared molecular ancestry establishes the basis for the use of animal models in medical research. Depending upon the system being investigated, the animal models may have to be quite similar to humans (an argument for the use of primates in research that focuses on infectious diseases), or they may not have to be similar at all (as is the case when the giant squid, *Loligo*, is used to elucidate neural function). The little lab mouse may indeed lose its position as the default laboratory animal model, but more likely it will just have to make room for other species to share its valued position. After all, outside appearances can be deceiving, as it is what's inside biochemically that truly counts.

Chapter 4

Chemical Journeys: Absorption

Everyone who got where he is has had to begin where he was.
— Robert Louis Stevenson

For a toxic chemical to be harmful, it must first travel from the environment to a specific target site within the body. Much of this journey happens through the deceptively simple process of diffusion: molecules migrating away from their source. Examples of diffusion are all around us, from salts dissolving into water to fumes dispersing from a smokestack at an industrial plant into the atmosphere. But diffusion becomes more complicated when the molecule must cross from one environmental compartment into another. For example, when a molecule diffuses from water into the air, properties of the molecule, air, and water all come into play to determine the rate at which diffusion occurs. Furthermore, the distance traveled by the molecule from its source and the dimension of the surface from which the molecule is diffusing also make a difference in the overall rate of diffusion.

For a toxic molecule to elicit a response in a target tissue, the toxic compound often has to cross several media. A pollutant in the air may have to cross from the air into a water layer (the moist environment within the lung), then from the water layer into a lipid layer (the cell membrane covering each living cell), then from the lipid layer back into the water layer within the living cell.

To reach the blood, the compound may have to repeat the effort again, moving from the watery interior of the cell, across a lipid membrane, and back into the watery realm of the bloodstream.

Regardless of how the chemical reaches an animal, whether the chemical is ingested, is inhaled, or enters through the skin, all chemical-biological interactions begin with movement of a chemical compound across an epithelial layer. The skin is the most obvious layer of epithelial cells, for it is what we see. The internal lining of the lung is a second layer across which transport can take place, while the lining of the stomach and intestines is a third. These three epithelial layers are not all the same, but differ depending upon their function.

Skin

Mammalian skin acts as a barrier between the outer environment and our internal ocean of blood. Our skin stretches tautly over our body, with only a few wrinkles or invaginations that only marginally increase its overall surface area. Even though our skin has a large surface area relative to the inner organs such as the liver, it is only as large as it needs to be to stretch over the rest of our bodies.

The outermost layer of skin is dead. The underlying dermal layer of the skin is continually dividing and producing cells that die, dehydrate, and keratinize. The process of keratinization involves the buildup of keratin in the epithelial cell layer. Keratin is a fibrous protein that is found not only in the upper levels of skin but also in nails, hooves, and even rhino horn. Bundles of keratin filaments are tough and, equally important, are insoluble in water. Consequentially, you can consider the keratinized layer of your skin to be, for the most part, a watertight boundary between you and the outside world.

Dead cells have no living protein associated with them; therefore the active transport of compounds across the skin cannot occur. But what about compounds that are fat-soluble? Since a dead cell still contains the lipids that were in the cell when it was living, fat-soluble molecules do absorb into the dead cell layer. The skin was not evolutionarily designed to act as an absorptive surface, but given that the protective barrier of the skin is made up of dead cells, the barrier by its very nature remains absorptive to lipid-soluble compounds.

The capacity of the skin to absorb lipid-soluble chemicals is both a blessing and a curse. Transdermal medicated adhesive patches, for example, take advantage of the capacity of the skin to absorb small lipophilic compounds in order to administer chemicals such as scopolamine, the anti–sea sickness drug, and nicotine. Unfortunately, urushiol, the active ingredient found in poison ivy, poison oak, and poison sumac, also takes advantage of the skin's capacity to absorb small, lipophilic chemicals, and is absorbed via the dermal pathway.

The Lung

While the lung is fundamentally different from the outer skin in the fact that its cells are all very much alive, the two do share one important characteristic. Namely, there is no protein-mediated transport of compounds from the air sac or alveoli into the lung. Beyond that, the similarities between the two epithelial layers pretty much end.

The lung represents a volume of space across which oxygen and carbon dioxide diffuse. In an amphibian such as a frog, the lung is configured like a bunch of grapes with a few very large air sacs, a relatively small total surface area (the combined surface area of all of the grapes in the bunch), and a relatively large diffusion distance from the center of the air sac to the animal's blood. In a mammalian lung of the same volume (the volume of a large bullfrog's lung is similar to that of a small rat), the alveoli are much smaller and more numerous. The rat also has a much higher metabolic rate, which necessitates faster diffusion of oxygen into its blood. This is partially accomplished by decreasing the diffusion distance (the size of the alveoli are much smaller) and by increasing surface area. Consequently, oxygen is delivered from the air to the blood, and carbon dioxide is removed from the blood to the alveoli, more quickly in the rat relative to the frog.

But it is not just the simple gases, such as oxygen and carbon dioxide, that diffuse across the lung epithelium, for vapors can also be absorbed. A vapor is the gas phase of a substance as it evaporates or volatilizes off of a liquid (think about vapors rising from a small drop of perfume), and all chemicals to do not generate vapors to the same degree. For example, water-soluble compounds do not volatilize, do not create vapors, and remain stubbornly dissolved in water. Even if the water droplet completely evaporates, these compounds are unlikely to enter the atmosphere, but rather will be left behind as a solid residue, often in the form of salts. Consequently, carrier proteins for the transport of water-soluble compounds are not found within the lung, as those compounds rarely present themselves to the lung in any appreciable concentration.

But what about the more volatile organic compounds? Even for these compounds there are differences in the degree to which they are absorbable, and the determining factor is the compound's blood:gas partition coefficient. To understand how the blood:gas partition coefficient governs absorption, consider a vapor enclosed in a cocktail shaker with a bit of water. After the cocktail is shaken, the vapor can partition either into the water or it can primarily remain in the atmosphere within the shaker. In this case, chemicals that remain in the shaker's atmosphere have low blood:gas partition coefficients, whereas chemicals found predominantly in the liquid have high values for the coefficient. These tendencies dramatically influence the capacity for absorption, because low values of the blood:gas coefficient are indicative of low rates of absorption, whereas elevated coefficient values predict much higher rates of absorption across the lung epithelium.

The fish gill is an interesting twist on a respiratory organ. Unlike the lung, the fish gill interacts with water, allowing for the uptake of dissolved oxygen from the water and the release of carbon dioxide from the blood directly into the water. In addition, the direct association between the gill and the water allows for the transport of water-soluble compounds. As such, the fish gill, unlike the mammalian lung, is enriched with proteins that allow for the transport of water-soluble compounds, specifically inorganic ions such as sodium, calcium, and potassium. The fish gill is not just a respiratory organ (an organ used to exchange oxygen and carbon dioxide with the aqueous environment), but is also an ionoregulatory structure, like the mammalian kidney, working to maintain appropriate concentrations of key elemental ions within the blood.

The Digestive Tract

While the chemical traffic across the skin is minor, and the traffic across the lung is restricted to a few highly specific chemical classes, chemical traffic across the epithelial layer of the digestive tract is fast and furious. The primary purpose of the gastrointestinal epithelium is to absorb food on a molecular level. Therefore, the intestinal epithelial membrane is festooned with proteins that function to move water-soluble compounds out of the intestinal tract and into the blood, where chemicals can be transported to the liver.

The transport of lipids across the intestinal epithelium, which might be expected to be simple, is actually quite complicated. When consumed, lipids—fats and oils—tend to mix together in the stomach and intestine, forming large globules, rather than remaining isolated as individual chemicals. These larger globules of fat have to be broken down into smaller droplets, a process known as *emulsification*. Emulsification occurs through the action of bile salts and acids that reduce the surface tension of the lipid globule, causing it to break into smaller globules called *micelles*. These small aggregates of fat and oil can then migrate across the digestive epithelium membrane, ultimately reaching the blood.

The chemical thoroughfare across the digestive system, so necessary to bring food molecules into the body, is also well designed for the absorption of toxic compounds. Proteins designed to transport water-soluble food molecules can also mistakenly transport water-soluble toxic compounds. These lipid-soluble compounds generally become incorporated into the large fat droplets within the intestine, and are also proportioned into micelles as they form. Along with the beneficial fats and oils within the micelle, the lipophilic contaminants ride the micelles across the epithelial membrane into the blood.

In the Blood

Once a chemical has gotten inside the body, it cruises through the blood vessels, like a gondola traveling the canals of Venice, toward its ultimate site of action—its target tissue. How chemicals travel through the vascular system depends upon a number of factors, including the chemical's solubility. Water-soluble compounds dissolve into the blood plasma and ride the flow, for the most part, in a free and unbound form. Lipid-soluble compounds, however, bind to proteins and an equilibrium develops between a small pool of free compound in the blood and a much larger pool of bound compound. This is important, as the free form is the biologically active one, being able to diffuse from the blood to the target receptor. But as the free compound diffuses from the blood into the extracellular fluid and the waiting target cell, the equilibrium shifts and more of the bound form is freed. The bound form of the compound represents a time-release capsule, slowly releasing free, and biologically active, compounds into the blood, where they can diffuse across the nearby capillary epithelium.

While a toxic compound may initially enter the body through the skin, lungs, or digestive track, its target is often in distant tissues. To get there, the compound will travel within the bloodstream. Its movement into the blood from the epithelial cells where absorption took place will be consistent with the process by which it made its way into the epithelial layers. For fat-soluble compounds, entry into the blood will not be an issue—they will not be barred by the cells making up the blood vessels. Water-soluble compounds, on the other hand, may have to use carrier proteins to shuttle them across cell membranes.

A compound's ability to move from tissues into the blood is not solely determined by its chemical attributes, for the blood vessels themselves either facilitate or frustrate this exchange. Blood capillaries within the brain adhere tightly together so that no transport can occur without the compound being directed through the cells of the capillary network. Consequently, water-soluble molecules can only enter the brain through the cells that make up the blood–brain barrier. In contrast, the capillary network in the liver is less densely organized, with holes or fenestrations so that bulk fluid can flow from the blood into the liver tissue and back again. The system allows for water-soluble food molecules (such as sugars) to enter the blood readily, but it also allows for water-soluble toxic molecules to follow the same route.

Once the compound enters the blood, its residence within the bloodstream depends upon its solubility. In water-soluble compounds, the chemicals remain dissolved in the aqueous blood and are trapped until they are allowed to exit, either via channels or gates, or through large fenestrations such as those that occur in the liver or the kidney. Water-soluble compounds are ushered around the bloodstream in a very controlled fashion. In contrast, lipid-soluble compounds enter, and often exit, the bloodstream in an uncontrolled fashion.

On a cellular level, our bodies wage an energetic, ongoing battle between control and chaos, and uncontrolled lipophilic compounds present a direct confrontation to our command over our internal environment. Fortunately, fat-soluble compounds are not maintained in their freestyle, go-anywhere form while in the blood, but rather are harnessed by binding to large and charged proteins. This binding creates a charged supermolecule, or protein–toxic compound conjugate that is essentially polar and therefore locked into place within the blood. For many toxic compounds (and nonpolar, nontoxics alike, such as sex steroids), a large pool of these conjugates are in equilibrium with a much smaller pool of free nonpolar compounds. As the free compound dissociates from the plasma protein, it resumes its nomadic life and can easily diffuse across the cell membranes of the capillary beds, leaving the blood. Therefore, the conjugated compound in the blood slowly liberates free compounds that can diffuse and become available to the waiting target receptors, where it can cause adverse impacts.

Sequestration

From the blood, the chemical compound can migrate to cell membranes and bind to target molecules on the cell surface, or it can enter the cell and bind to the target molecules within. When a compound enters a target cell, it can follow a number of different routes, with various consequences beyond the obvious one of damaging the cell. Some compounds can become sequestered—that is, embedded in the body in a stored form, remaining relatively benign. For example, water-soluble metal ions can become incorporated into bone, whereas lipophilic compounds can become incorporated into fat deposits.

Sequestration is well illustrated by the mystery surrounding Napoleon Bonaparte's death in 1821. Bonaparte's last will and testament avowed: "I die prematurely, assassinated by the English oligarchy and its tool. The English nation will not be slow in avenging me." The passage led to speculation that the great leader had been poisoned. Fortunately, his staff had kept locks of his hair, and when it was tested for arsenic content, it registered 100 times the normal level. While the arsenic sequestered in Napoleon's hair may not have been his undoing, the chemical analysis did provide unequivocal evidence that Napoleon had been exposed to arsenic during the latter days of his life.

Napoleon's story has recently taken an unexpected turn: up-to-date studies have revealed that arsenic poisoning was probably not the culprit in this who-done-it. He is now believed to have died of peptic ulcer and historic gastric cancer. Interestingly, wallpaper from

Longwood House on Saint Helena Island, where Napoleon lived in exile, had a particular green floral pattern. Wallpaper of that time was made vivid green by using copper arsenite (Scheele's green) in the paint. When copper arsenite becomes damp (not unlikely, considering that the house was on an island), it is converted by molds into trimethylarsine gas, and it is most likely that it was this gas, not an assassin's poison, that was the source of the arsenic found sequestered in Napoleon's hair.

Compounds that are sequestered can remain in tissue depositions for life, or they can be removed from the body in an inert form (such as the arsenic in Napoleon's hair), or they can be liberated back into the bloodstream. Regardless of the specific pathway along which the chemicals travel or their level of toxicity, chemicals obey the rules of diffusion and are carried toward their designated target tissues accordingly.

Chapter 5

Bodily Defense

If open country stretched before him, how he would fly, and indeed you might soon hear the magnificent knocking of his fists on your door. But instead, how uselessly he toils; he is still forcing his way through the chambers of the innermost palace; never will he overcome them; and were he to succeed at this, nothing would be gained: he would have to fight his way down the steps; and were he to succeed at this, nothing would be gained: he would have to cross the courtyard and, after the courtyard, the second enclosing outer palace, and again stairways and courtyards, and again a palace, and so on through thousands of years; and if he were to burst out at last through the outermost gate—but it can never, never happen—before him still lies the royal capital, the middle of the world, piled high in its sediment.

— Franz Kafka, "A Message from the Emperor"

Given the enormous importance of the human brain, one might think that our bodies would have evolved extensive safeguards to protect it from toxic compounds. Yet the brain is so easily intoxicated with some very common chemicals that we don't even question it. Alcohol, nicotine, caffeine, heroin, methamphetamine, cocaine—they all travel in the blood, reach the cranium, and enter into the cerebral spinal fluid so quickly and efficiently that the brain seems to be defenseless against chemical onslaught. If the brain is so critical to our existence, how is it that the psychoactive substances mentioned above, which are all toxic molecules, can all so easily enter into the inner recesses of the brain and alter its function?

Ironically, while it is fairly straightforward to intoxicate oneself by taking psychoactive drugs, delivering therapeutic drugs to the brain is much more difficult. Blood capillaries in the brain are composed of endothelial cells that are very tightly joined together. Therefore, any chemical transport from the blood into the cerebral spinal fluid must occur through the cell, as there are no gaps in the capillary through which the chemicals can move. All of the psychoactive drugs mentioned previously, from caffeine to methamphetamine, are lipophilic molecules; they do not see the endothelial cell as a barrier but rather pass easily from the blood into the brain, where they can elicit their effect on nerve cells. In contrast, many pharmaceutical drugs are water-soluble, and hence the blood–brain barrier, as the endothelial layer is called, poses a very real challenge to therapeutic drug delivery.

Whether focusing on the brain or any other tissue, humans (and animals) are not passive victims of chemical insult. Rather, they have evolved mechanisms that can reduce, and in many cases, completely stop the inflow of toxic chemicals. But the effectiveness of these defenses depends on both the chemical and its target. In fact, the very traits that facilitate the transport of some chemicals toward a biological active site are effective pathway barriers for others. Cell membranes themselves, from the epithelial layer of the blood–brain barrier to the tough, cornified outer epithelial layer that makes up our skin, are incredibly effective barriers against water-soluble toxic compounds but are relatively ineffective barriers when it comes to fat-soluble toxic compounds. In this chapter, we will examine the various defenses the body has evolved to protect itself from toxic compounds, as well as ways that those protections are breached.

Misdirection and Sequestration

A chemical's transport from the environment to a biological active site can follow any of several alternate pathways, or side routes. If the molecule happens to reach a metabolically active tissue, such as a neuron or a liver cell, it may do substantial harm. If, however, the toxic compound entering the body can be misdirected or sequestered into a relatively inactive tissue, such as bone or adipose tissue, then the toxic compound may not reach receptor sites in more metabolically active tissues, at least not in the quantity necessary to elicit an immediate toxic impact. Metal ions sequestering into bone or hair and lipid-soluble compounds such as DDT sequestering into fat are two classic examples of such containment.

If a chemical compound can make it past these outer lines of defense and actually enter a target cell—for example, a nerve cell in the brain—the journey is not yet complete. Within the cell there are also a number of additional lines of defense that further impede the progress of the toxic compound as it migrates toward its target destination.

One internal defense that a cell can muster against a toxic insult is the production of alternate binding sites. Again, the overall strategy is sequestration or misdirection. The greater the number of molecules of the toxic compound that bind to these alternate binding sites, the fewer that remain free to bind with the cell's target site and cause damage. A classic example of an internal sequestration site within animals' cells are *metallothioneins*. Metallothioneins are fairly small, cysteine-rich proteins that aggressively bind to a number of metal ions. Metallothioneins may have evolved to bind aggressively with metal micronutrients, such as copper and zinc. These compounds are fairly rare in the diets of most animals (particularly plant-eating herbivores), and an internal storage depository for these ions may have been evolutionarily advantageous. While these micronutrients are essential at low levels, they are also toxic at higher levels (see chapter 1). Furthermore, other nonessential metal ions, such as cadmium, silver, and mercury, also bind to metallothioneins, thereby reducing the free concentration of the ions in the cell and thus reducing the overall toxicity.

Biological Barriers

Up until this point, we have examined defenses against toxic substances that involve physical barriers, misdirection, and sequestration. Within the cell, however, metabolic machinery can come into play and directly modify the structure of the toxic compound. This process, known as *biotransformation*, is a key line of defense by which toxic compounds can be deactivated. For some contaminants, biotransformation can break the molecule down into smaller and less toxic components. For example, alcohol is completely metabolized to carbon dioxide and water, and is therefore not likely to persist in the brain of the imbibing individual. Not all compounds, however, are equally susceptible to biotransformation. In contrast to alcohol, most compounds containing one or more elements from the halogen column on the periodic table—namely flourine, chlorine, bromine, and iodine—are highly resistant to biotransformation. Since many of these compounds are lipid-soluble, they can become sequestered in fat, where they can remain for years to decades. Some of the better-known halogenated compounds include the pesticide DDT and the polybrominated diphenyl ethers (PBDEs) used as flame retardants. (The subject of halogenated compounds will be discussed in greater detail later in the book.)

A subset of all biotransformation processes, and an important one in toxicology, involves the creation of a water-soluble compound from a lipid-soluble one. While this may not deactivate the toxic compound, it does allow for it to be excreted from the organism via the kidneys or the intestine. Two broad classes of proteins perform this conversion in processes known as *Phase I* and *Phase II metabolism*. The name is a bit unfortunate, as Phase II metabolism does not necessarily follow Phase I, nor do both processes have to happen in sequence before an excretable metabolite is created. When metabolized, the parent compound (the one that has migrated into the cell from a distant location) is converted into one or more metabolites that are, generally speaking, more water-soluble than the parent. Furthermore, in most but not all cases, biotransformation reduces the toxicity of the parent compound.

Phase II metabolism is both predictable and specific. The proteins involved in it are highly selective catalysts that facilitate the conversion of chemical substances known as *substrates* into other chemicals known as *products*. Phase II enzymes are highly selective for their substrates and only catalyze the production of very specific products. The range of diversity in the action of these enzymes is limited, as they generally elicit their action on a specific group of substrates and thus generate, though contortions of the protein, a relatively small number of products. Their primary function is *conjugation*, which is to say that they attach a major side group onto the parent compound, thereby making it polar or water-soluble. By making the molecule more water-soluble, this process also makes the molecule excretable in the urine. Furthermore, the Phase II proteins are generally *cytoplasmic*, meaning that they bob around in the watery, protein-rich soup within the cell, rapidly and opportunistically conjugating fat-soluble molecules as they enter the cell.

In contrast to the Phase II enzymes, Phase I enzymes are the generalists of the protein/enzyme world. The products of Phase I metabolism are less specific, as many Phase I enzymes act on a broad range of molecules—their substrates—and they also have the capacity to generate a wide variety of products. The primary function of a major class of Phase I enyzmes, the cytochrome P450s, is called *monooxygenation*, as these enzymes attach part of a water molecule (the OH or hydroxy portion of a water molecule, H_2O) onto one of a number of potential binding sites on the parent compound. This process only moderately increases the water-solubility of the parent compound, but that may be enough to allow it to be excreted. Rather than being found within the cytoplasm of the cell, these enzymes are embedded in the inner membrane of mitochondria, and in fact, Phase I metabolism involves not one but two proteins that pass electrons from one to another. A consequence of this cellular architecture is that the overall rate of reaction tends to be slower than that of the Phase II enzymes.

When Phase I and Phase II metabolism occurs in tandem, it can sometimes lead to a smooth conversion of a lipid-soluble compound into a water-soluble one. However, for some chemicals the interplay between the two systems can sometimes also lead to unintended consequences. A perfect example is the metabolism of acetaminophen, the active ingredient in Tylenol. The general user may believe that Tylenol is totally harmless and that "if a little bit is good, then more is likely to be better." This is dangerous misconception. The elimination of acetaminophen from the body falls under the domain of a few Phase II liver proteins. Generally, the metabolic pathway of acetaminophen, during biotransformation, is such that a water-soluble conjugate (the acetaminophen skeleton bound to a sulfate or a glucuronide side group) is formed. However, during that formation, about 2 percent of the parent compound is converted by Phase I metabolism into two different metabolites (known as *quinoneimines*). These molecules are highly toxic, causing kidney and liver damage, but fortunately they are rapidly converted (by still other Phase II enzymes) in the liver and kidney into a conjugated, water-soluble, and excretable metabolite. As such, the quinoneimines are short-lived and are rapidly converted into nontoxic conjugates.

During acetaminophen overdose, however, the biotransformation pathway for the parent molecule remains the same, even as the amounts of both the conjugate and the quinoneimine within the cell increase dramatically. Consequently, the toxic effect is greatly enhanced. If Tylenol is combined with alcohol, the production of the quinoneimine is similarly increased, as the alcohol simultaneously enhances the activity of the Phase I enzyme (thereby increasing the production of the toxic metabolite), while also decreasing the activity of the Phase II enzyme (thereby slowing the rate at which the toxic metabolite is converted into a conjugated and excretable form).

This example illustrates a number of important aspects related to biotransformation. The first is that Phase I metabolism does not inevitably lead to Phase II metabolism but can occur separately from it. The second is that the metabolites of Phase I metabolism can be more toxic than the parent molecule! Here is a classic biological example of the adage "Let no good deed go unpunished," as the efforts of the cell to render the molecule more water-soluble, and therefore more excretable, can also enhance its toxicity. Thus, biotransformation is not a sure-fire panacea for the elimination of all toxic organic compounds, but nevertheless it is the best metabolic defense for the elimination of lipid-soluble toxic compounds that has evolved to date.

If a chemical has eluded all the previously described defenses, the cell may still have one last strategy to thwart the docking, or binding, of the toxic compound with its long-sought-after target site. Some sites require a very precise docking between the toxic molecule and the receptor site. In this situation, very slight changes in the structure of the receptor site, or lock, may prevent the chemical, or key, from fitting inside. These alterations can occur if the target is a protein. Since proteins are made up of individual amino acids strung together and then twisted into a three-dimensional structure, small changes in that structure can dramatically alter the protein's shape, and thereby its function. Alterations in the amino acid structure of the receptor protein have to be approached (from an evolutionary perspective) with caution, as large-scale changes in the receptor site may interfere with its intended function, along with its reduced proclivity for toxic chemicals.

Paracelsus Revisited

The rules of Paracelsus, our old friend from chapter 1, help to put the body's defensive mechanisms into a more robust context. The magnitude of the impinging wave of toxic molecules, in most circumstances, directly influences their effect. At low exposure doses, the number of toxic molecules interacting with biological receptors is so few that no adverse impact develops. This can be due, in part, to an insufficient number of molecules binding to the receptor of the target site, but it can also be due to the effectiveness of the defensive mechanisms impeding the flow of molecules toward the receptor site. At higher exposure doses, the increased number of toxic molecules migrating toward the receptor sites increases the number of molecules that actually run the defensive gauntlet and reach the target site. At still higher concentrations, the defensive mechanisms can be overrun, the number of toxic molecules reaching the receptor site can increase dramatically, and the adverse impact, such as that which occurs during an acetaminophen overdose, can be profound.

Chapter 6

Wider Journeys: Pollution

Suddenly, from behind the rim of the moon, in long, slow-motion moments of immense majesty, there emerges a sparkling blue and white jewel, a light, delicate, sky-blue sphere laced with slowly swirling veils of white, rising gradually like a small pearl in a thick sea of black mystery. It takes more than a moment to fully realize this is Earth.

— Edgar Mitchell, *Apollo 14* astronaut

Proponents of the Gaia hypothesis maintain that the biosphere (the part of the Earth's surface and atmosphere that supports life) is a single, self-regulating system that maintains conditions for life on Earth. While the point of this chapter is neither to refute nor to support the Gaia hypothesis, the concept of the Earth as a single system has a great deal of traction within the field of toxicology. In fact, there are a number of strong parallels between the movement of chemical compounds across the Earth's surface, and the absorption, delivery, sequestration, and biotransformation events that occur within individual animals.

The Three Compartments: Air, Water, and Land

Earth's natural environment can be divided into compartments, much the same way that an organism can be divided into different organs or organ systems. For the biosphere, three of the most relevant compartments are the atmosphere, the water, and the landmass. When toxic and nontoxic molecules are released into the environment, they can move from one compartment to another, with their movements governed by attributes of the chemical as well as attributes of the environment in which it is moving.

Changes from one compartment to another are known as phase transitions, as compounds can transition from solid to liquid to gas. For elements and small molecules, these transitions are intimately associated with geochemical cycling as elements naturally transition from one phase to another. The common elements carbon, nitrogen, and phosphorus have the vast bulk of their matter locked up in solid form within the crust of the Earth, where they may remain stable for eons. A much smaller pool of these molecules may transition from solid to being dissolved in water, to being volatilized into the atmosphere.

The best-known, and most obvious, geochemical cycle is that for water, also known as the hydrologic cycle. Water in the ocean can evaporate and enter the atmosphere, migrate across large distances, then return to liquid water in the form of rain, which can then flow across the landscape, ultimately reaching the ocean where it began its journey. Not all of the chemicals are as transitional as water, and many elements do not readily transition into the atmosphere as gases, as they are not capable of vaporizing. Nevertheless, many elements undergo phase transitions as they move back and forth from being locked in a solid to being dissolved in water.

The chemical cycles, while natural, are not necessarily benign. For example, arsenic can leach naturally from rocks and dissolve into groundwater. The concentration of arsenic in that water may be high enough to be toxic to humans, rendering it useless for drinking unless treated. Likewise, mercury can transition from earth to atmosphere during volcanic eruptions, ultimately depositing in aquatic systems where it can similarly be toxic.

Larger chemicals, whether natural or synthesized by humans, can also enter the environment and, if persistent enough, they too will begin to cycle, transitioning from one phase to another. These chemical compounds may undergo phase transition from solid to liquid to gas, depending upon their specific properties. Of course, chemicals, unlike elements, rarely maintain their chemical form during these phase transitions. Often, abiotic processes such as photooxidation or combustion may alter their chemical structure. Regardless of their ultimate structural fate in the environment, the cycling of larger chemicals depends upon their chemical characteristics, such as the ability to ionize when immersed in water or their ability to vaporize when exposed to the atmosphere.

Organisms: The Fourth Compartment

When viewed as a composite, the sum of the Earth's biota (microbes, plants, animals) can be viewed as just another compartment through which elements and chemicals cycle. For example, when an animal drinks water, eats food, or inhales air, it inadvertently becomes part of the geochemical cycling of the chemicals within those substances. Once those elements and chemicals are released back into the environment, they are capable of rejoining the inanimate geochemical cycle.

Relative to the other three compartments, however, organisms dramatically increase the overall rate and complexity of chemical cycling. Consider the geochemical cycling of calcium. Exposure of calcium-bearing rocks to water allows the calcium to become ionized and released into solution. The ionized calcium (Ca^{2+}) reacts with dissolved carbon dioxide (CO_2) to form calcium carbonate, or limestone, which settles out in solid phase, where it can remain for hundreds of millions of years. Animals have exploited this comparatively simple phase transition from ionized calcium in water to the production of calcium carbonate, and they have incorporated the salt into some of the hardest biological tissues. Bone, turtle shell, and the shells of clams and other invertebrates all contain calcium carbonate. Furthermore, ionic calcium is an important cell signal involved in muscle function and the activation of intracellular proteins. In fact, in most organisms with bone or shell, the calcium carbonate is in equilibrium with that in the blood (the liquid compartment). The exchange of calcium ions between the two is continuous, as the residence time within an organism is at most on the order of decades.

As illustrated in the calcium example above, the time that a chemical or element is trapped within an organism is trivial when compared to the time course of many of the other components of geochemical cycles, such as the gradual breakdown of rock formations during freeze-thaw cycles or other hydrologic events. Therefore, the slow geochemical cycling that occurs in abiotic environments stands in stark contrast to the frenzied, though brief, kinetics of the ion when in a biological system.

For an organism to become directly involved in the geochemical cycling of an element, such as calcium, or a larger molecule, such as a steroid released in the effluent from a wastewater treatment plant, the compound has to be *bioavailable*. The concept of bioavailability arose from the field of pharmacology, and by definition a chemical injected into the bloodstream is 100 percent bioavailable. The important point here is that a chemical can be considered bioavailable if it can be absorbed from the environment into the animal's blood. Bioavailability differs from bioactivity, in that a *bioactive* compound needs to reach the site of physiological or biochemical activity where it can elicit a biological effect.

Bioavailability can be either an active or a passive process. Consider for a moment a pond filled with organisms (fish, algae, microbes). If a lipid-soluble molecule is added to the pond, there will be a strong propensity for it to enter the lipid components of the environment—in other words, the living organisms. The partitioning of the lipid-soluble compounds into the animals will be passive, as the compounds will be able to absorb fairly easily into the aquatic biota. Alternatively, if there are metal ions in the water, they will not be able to enter into the biota within the pond without help, due to their polarity and the reticence of polar compounds to cross the lipid bilayer, as previously discussed. In this case the absorption of the compound into an organism would have to be active, and it would involve the help of protein-facilitated transport, as opposed to simply being a passive process.

Whether toxic or not, compounds that ionize in water are almost always bioavailable to organisms to some degree. The question is not whether the compound will enter the resident biota, but rather, the rate at which the absorption will occur. While the lipid bilayer presents a barrier to the uptake of ions from the water, there are enough ion gates and pores within the cells of the gill or intestine of an aquatic animal that transport of ions into the blood is occurring continuously. Therefore, bioavailability of most ions is ubiquitous. Likewise, lipophilic compounds also tend to be ubiquitously bioavailable regardless of whether the compound is volatile and therefore available through atmospheric inhalation, or nonvolatile and available only through direct ingestion or via absorption across the skin.

Living organisms, therefore, can be considered just another compartment through which elements and chemicals transition. Yet unlike the other compartments, biological systems are active participants in the transitioning, as they can redirect the flow of chemicals from one internal compartment to another, or from the inside of the organism back out to the environment.

If the rate of entry of the chemical into the the resident biota exceeds the rate at which it can be excreted, two eventualities occur. The first is that the chemical can be transformed into another one that can be dealt with more easily. On the other hand, if the chemical cannot be transformed, then it will accumulate with the organism's tissues, increasing the bioconcentration of the compound. *Bioconcentration*, like bioavailability, has a very specific definition, as it is the process by which a chemical directly enters into aquatic organisms from water. Often it is presented in terms of a *bioconcentration factor*, which relates the concentration of a chemical in the organism relative to the concentration of the chemical in the water. For chemicals that tend to be lipophilic, bioconcentration factors can easily be greater than 1, meaning that the chemical is present in greater concentration in the organism than in the water.

When an animal consumes another organism (plant, animal, or bacteria) and the compounds within the food item are incorporated into the tissues of the consumer, the trophic increases in chemical concentration are known as *biomagnification*. Biomagnification has often been identified in aquatic systems, and once again the lipophilic and persistent organic compounds represent an important class, though not the only one, of compounds that bioconcentrate. Phytoplankton, the small photosynthetic plant cells found in lakes, ponds, and oceans, are lipid-rich cells that can act to bioconcentrate fat-soluble compounds from the water. In turn, when crustaceans called zooplankton consume the phytoplankton, they further bioconcentrate the fat-soluble compounds. This cascade continues up the trophic levels, such that the small fish consuming the phytoplankton, and the larger fish (or mammals or predatory birds) consuming the smaller fish will continue to biomagnify the compounds.

A specific historic case regarding biomagnification of a persistent organic pollutant involves the repeated spraying of DDT on Clear Lake, California, the largest natural freshwater lake found entirely within the state's borders, to kill the larvae of the Clear Lake gnat. The gnat resembles a mosquito, without the needlelike biting apparatus, and poses no threat to humans, beyond the indelicate specter of inhalation. Normally, the gnat's larvae develop in water and begin to emerge from the water in the spring after March. Before the use of pesticides, the gnats were so pervasive that piles of dead insects appeared in masses beneath streetlights; furthermore, any motorists visiting the lake during the summer could be greeted with swarms of gnats so thick that they would have to repeatedly stop every quarter mile or so to scrape the gnats off their windshields and headlights. From 1948 through 1957, the lake was repeatedly sprayed with DDT and the gnat population was dramatically reduced. These sprayings also led to waves of mortality that wiped out a breeding colony of western grebes. Bioconcentration of DDD (a metabolite of DDT) was clearly shown in the food chain of Clear Lake. A composite sample of phytoplankton contained 5.3 parts per million DDD, which was over 250 times greater than the concentration in the lake water. Small fish contained twice the concentration found in the plankton. When scientists determined the concentration of DDD in the breast fat of the western grebes, fish-eating birds at the top of the lake's food chain, they found levels up to 85,000 times higher than that of the lake water.

When animals, including humans, are involuntarily exposed to chemicals in the atmosphere or water, they unknowingly become a component of an abiotic geochemical cycle. Compared to the smoking of tobacco or the ingestion of alcohol, involuntary exposure doses and their associated health risks are much harder to quantify. Nevertheless, these exposures may cause real harm, which can become even more pronounced when the chemicals bioconcentrate or biomagnify.

Regardless of whether or not the Earth is a self-regulating system, it is clearly self-cycling, as chemicals move, often through geologic time, from one transition phase to the next. Like it or not, the life-forms on Earth are all along for the ride, and in the case of environmental exposures to air and water, living things are short-term resting places for chemicals as they move through their geochemical cycles.

Chapter 7

Traveling Particles

The wind blows to the south and goes around to the north; around and around goes the wind, and on its circuits the wind returns.
— Ecclesiastes 1:6–7

In the previous chapter, we explored the transport of chemicals from one physical compartment to another—but only in their free molecular form. The reality, however, is much messier than that, as solid particles can travel in water and both solid and liquid aerosols can travel in the air. While particles moving in water and those moving in air are governed by different forces, there are also some striking similarities. One is the relationship between particle size and travel distance: the smaller the particle, the greater the distance that it is likely to travel. The second deals with colonization, as toxic chemicals can bind to particles, hitching a ride on them as they travel downstream or downwind.

Solids Carried in Water: Sediments

If sediments are just solid particles of different sizes, the large particles are clearly the more charismatic. The snow-white beach sand of the Caribbean and the massive boulders of a Colorado mountain stream provide unparalleled photo ops, but they also give a skewed view of the interaction between water and sediments. Rather than two discrete environments that don't mix, sediments and water are the yin and yang of the aquatic environment, in that they both complement the whole. Water moves sediment particles, and the greater the discharge of water, the larger the particles that are moved. Likewise, sediments interact with the water, trading molecules back and forth from being dissolved in the water to being bound to sediment particles both large and small.

The largest sediment particles are known as *boulders*—chunks of rock that have diameters in excess of 25 centimeters (about 10 inches). Smaller chunks can be differentiated into *cobble*, *gravel*, *sand*, *silt*, and *clay*. The very smallest particles are *colloids*—particles so small that they remain suspended in the water even when the water is still. Colloids do not behave strictly as particles (in that they do not settle out from the water), nor do they behave as true water-soluble compounds (in that they can be filtered out of a solution). Colloids fall between the netherworlds of particles and water-soluble compounds. As is true with plasma proteins in the blood, colloids can influence the concentration of chemicals in water by binding with them, removing them from the dissolved phase, and thereby reducing their toxicity.

But if colloids remain in suspension, why aren't the oceans and rivers perpetually full of colloidal material? Two reasons: first of all, inorganic and organic colloids, under certain environmental conditions, will aggregate together, become flocculent, and settle out of the water, taking whatever chemicals that have bound to them to the bottom. Second, organic colloids are food, and these colloids will be collected and eaten by detritivorous animals and also broken down by ravenous microorganisms. For some toxic molecules, the hungry bacteria that eat the colloidal material can transform the toxic compounds from one form to another or break it down entirely. In some environments, colloids and toxic chemicals may be in a state of perpetual flux. In environments where colloidal material dominates, the toxicity of waterborne chemicals may be ameliorated, while in other environments where the colloidal material is largely absent, the toxic compounds may remain in solution and achieve the highest potency.

As sediment particles increase in size from colloids to clays and larger, their transit in moving waters requires more energy (meaning that the water has to be moving faster) or else they will settle out of solution. However, even when residing on the bottom of a waterway, sediments still play a vital role in the motility of toxic molecules. Part of the reason for their involvement is that small sediments, such as clays, both attract and repel chemical compounds. The surface of a sediment particle is generally negatively charged; therefore negatively charged chemicals in the water will be repelled from the particle, whereas positively charged molecules will adhere to it.

Generally speaking, the smaller the sediment particle, the more important its role in the movement of chemicals through the aqueous environment. There are at least two important reasons for this, the first being the particle's surface-area-to-volume ratio. To understand this, consider puncturing the skin of a basketball and laying the skin out flat. In this case, the skin would cover a certain amount of area, roughly the dimension of a small hand towel. However if you were to fill the volume of the basketball with marbles, then spread out the surface area of those individual marbles, the surface area would approach the size of a small bedsheet, or roughly ten times the surface area of the basketball. The surface of all of the particles within a set volume is known as the *specific surface area*, and is important as it explains how a handful of pebbles can contain many more binding sites for chemicals than can a single large rock. While it is a bit more difficult to envision, the same is true for the smaller particles; a spoonful of clay or silt contains many more binding sites for organic material than does a spoonful of sand.

The specific surface area of differently sized particles explains a lot about the interaction between chemicals and sediments, but it is not the full story. As the particle size decreases, the space between particles also decreases (consider the size of the space between stacked basketballs and the size of the spaces between stacked marbles). In aquatic systems these spaces are filled with water, which is specifically known as *pore water*—the water in the spaces, or pores, between sediment particles.

In mountain streams, the stream water is moving quickly and the flow of water through these pore spaces washes out silts and clays along with any organic material, such as bits of decaying leaves or grasses, leaving behind an environment of clear water, large substrate, and little organic material. Toxic chemicals residing within the pore water, or on the small bits of organic material within that space, are more likely than not to move downstream in the current rather than interact with the minimal sediment that is there. In contrast, waters in a lowland stream, such as a stream meandering through a cornfield in Illinois, are moving relatively slowly and small particles (sands, silts, and clays) are being deposited on the river bottom. Any organic material that is moving downstream may also become deposited in the pore spaces between these small inorganic particles. That organic material is much more lipophilic than the inorganic material surrounding it; therefore lipophilic compounds in the water will be attracted to it. Consequently, while lipophilic compounds that adhere to organic material in mountain streams may move downstream, the same compounds are likely to move into the organic sediments of slower-moving rivers.

We intuitively understand these relationships between sediment particle size and bacterial activity, and it can be seen in our particle terminology. For large sedimentary particles (from gravel to sand) the name of the particle does not change when the material is wet or mixed with organic material. Wet sand is still sand, wet gravel remains gravel. In contrast, the nomenclature for silts and clays changes when these particles become wet and when they are mixed with organic matter. Wet clay and silt is known as mud, and mud containing a substantial amount of stable organic material is known as muck.

For most people, mud and muck evoke negative images, and for good reason. Bluntly, they stink. Muds and mucks are the site of decomposition (biotransformation on a macromolecular scale), and that decomposition gives off gases and vapors that can smell bad. The bacterial residents within these communities are the bacteria of decomposition and decay, necessary functions performed by the ecosystem, but ones that we would just as soon not think about.

As such, the colonies of bacteria that grow in muds and mucks are much more numerous, diverse, and vigorous than those that grow between grains of sand. All of these bacteria are on the hunt for food molecules, and once they find them, the bacteria will biotransform some of the adhering molecules, changing their form and, in some cases, rereleasing them into the water. It is not uncommon for one compound to enter the upper sediment layers at the bottom of a stream, and for a different metabolite of that compound to be released from those same sediments back into the water column.

Recent research has suggested that compounds released from organically rich sediments may carry a toxic wallop. In a recent study in Nebraska, sediment and water were collected from a number of local environments, brought back from the field to the laboratory, where a local fish, the fathead minnow, was subjected to the mixture with surprising results. The water with the sediment was collected from regions contaminated with agrichemicals, particularly during the spring just after pesticide application. When female minnows were exposed to a laboratory combination of river water and river sediment, the fish experienced the same adverse impact (defeminization) as did fish caged directly in the river itself. Unexpectedly, fish exposed to clean lab water over field sediment also experienced adverse impacts. Clearly, the sediment was not merely acting as a one-way sink for contaminants, but was also acting as a source of the agrichemicals.

Solids and Liquids Carried in the Atmosphere: Aerosols

The location of sediments on river bottoms and ocean floors precludes much direct interaction between deposited sediment particles and people. In contrast, particulates that are suspended in air, even temporarily, can be inhaled by us, causing health problems. As with sediment particles, airborne particulates are characterized primarily by their size, and their size influences their overall toxicity.

To think about the size of airborne particulates, consider the dust kicked up by vehicles on a dirt road. The particles that will become airborne are all less than 100 micrometers in diameter (for reference, a fine human hair is between 40 and 70 micrometers in diameter) and are collectively referred to as total suspended particulate matter (TSP). The largest of these particles (from 100 microns to 10 microns) tend to settle to the ground quickly and do not travel far from the dusty road.

Naturally, as particles get smaller they also get lighter, and consequently they remain suspended in air longer and travel farther from their source. If inhaled, these particles can continue to travel along the airways within the body, from the nose and mouth to the alveoli of the lungs. The journey to the lungs is a tortuous one, as the sinuses and the oral cavity contain twists and turns along the path. Coarse particles, those between 10 microns and 2.5 microns in diameter, tend to collide with the inner walls of the airways, sticking to the mucous lining and halting in their progress toward the alveoli. The fine particles, in contrast, will be able to ride the current of air rushing into the lung all the way to the alveoli.

With respect to air pollution, the United States Environmental Protection Agency focuses upon two particle classes: PM10 and PM2.5. Unlike the situation with sediment particles in water, the distinction between PM10 and PM2.5 particles is not really size, as the classifications are maximal particles sizes. In other words, PM10 dusts include particles that are 10 microns in diameter and smaller, including sizes of particles that are 2.5 microns and smaller.

Interestingly, for many PM10 dusts, up to 50 percent of the total mass of the particulates may be in the very small fraction, those less than 2.5 microns in diameter. Another distinction between sediments in water and dusts in the air is the importance of surface-area-to-volume ratios. Aerosols do not pack together as do solid particles; therefore the concepts of specific surface area and pore space between the particles offer little explanation for the toxicity of airborne particles.

The distinction between PM10 and PM2.5 particles is primarily based upon the aerosol's source and its composition. PM2.5 fine particles are principally the products of combustion, and common sources include the exhausts from cars, trucks, and buses, as well as the smoke from burning fuels such as wood, heating oil, or coal. These tiny particles can penetrate the lungs at the level of the alveoli, where they will dissolve into the aqueous film and actually become absorbed across the lung epithelium into the blood. These particles are so small that they begin to act effectively like macromolecules rather than very small particulates. Fine particles are also often associated with organic compounds that can be biotransformed, dramatically altering their toxicity (as will be discussed in chapter 10).

The PM10 classification generally holds for coarse particles, or what is known as "fugitive dust." *Fugitive dusts* are composed of granular bits, such as salt particles from evaporating seawater, pollen, spores, tire particles, and soil materials that are kicked up by driving on dirt roads. When inhaled, as stated previously, the larger particles may get trapped as they travel down the airway to the lung. However, given the fact that up to 50 percent of the mass of fugitive dusts are particles of less than 2.5 microns in diameter, the small-particle fraction within this dust will reach the inner recesses of the lung, where it can cause respiratory illness and lung damage.

Aerosols are much less likely than waterborne sediments to foster a microbial community. This is partly because there is no habitable "pore space" between particles in the air. Another reason is that most dusts and smoke present a dry environment that is relatively uninhabitable for microbes, compared with the wet surfaces of submerged sediments. This is particularly true for PM2.5 materials, as they are often the product of combustion, in which most bacteria would succumb to the flames. Despite these obstacles to colonization of aerosols, it does occur and a good example comes from the dusty plains of West Texas.

In Texas, the western high plains are notoriously dry and windy. The region is also known for its beef production, and when the cattle defecate, the waste dries quickly and is then trampled by the cows' hooves. The combination of a hot and windy environment leads to rapid desiccation, and the mechanical pulverization creates ideal circumstances for the development of fugitive dusts that travel downwind from the feedlots. Fecal particulates, veterinary pharmaceuticals, and steroidal growth-promoting compounds all become attached to dust that is blown from the surface of the feedlot. In addition, the dusts can carry substantial amounts of fecal bacteria. In this manner, potentially toxic chemicals and virulent bacteria can travel downwind from feedlot operations to where they may cause pollution problems.

On a clean and simplified Earth, contaminants would remain in one form unless they could convert wholly to another. The interactions between solid, liquid, and gas would be relegated to the relatively straightforward processes of freezing, melting, boiling, and sublimation. The ecosystem, however, is not so clean or simple. Solids can travel in the water or air as particles, governed by unique processes that do not come into play when solids dissolve into liquids or vaporize into the atmosphere. Dusts and suspended sediments act as vehicles by which toxic compounds can travel, sometimes great distances, in wind or water.

Chapter 8

Toxins, Poisons, and Venoms

Bee stings are very educational.
— Garth Nix

Today, when we think of toxic substances, synthetic chemicals tend to come to mind. But in order to understand human-made toxic compounds, technically known as *toxicants*, it is helpful to first look at *toxins*, those toxic compounds that occur naturally. Naturally occurring toxins can be further segregated into *poisons*, the chemicals that enter the body via contact between host and victim, and *venoms*, the chemicals that are injected into the victim's body via stingers, fangs, or teeth.

Eating Plants and Animals

There are over 400,000 different plant species on Earth, and within them all is a dizzying array of organic chemicals. Given the number of chemicals available for consumption among plant species, it is not surprising that some plants contain chemicals that are poisonous to the animals or the people who eat them. Indeed, even plants that we generally consider safe for consumption can contain toxins that are found either in parts of the plant that we don't eat or in the edible portions of the plant when it is not properly prepared. For example, chemicals known as cyanoglycosides are found in apple seeds as well as peach and cherry pits, and when they come into contact with the enzyme β-glycosidase, the result is a highly toxic chemical, hydrogen cyanide. Many fruit trees contain cyanoglycosides in their leaves and seeds, with only negligible amounts in the fleshy part of the fruit. However, cassava root, a major source of carbohydrates for over 500 million people worldwide, contains considerable amounts of cyanoglycosides within the edible root itself. It is widely understood that the tuber has to be properly processed to reduce the cyanide to nontoxic levels prior to consumption.

Ingestion of plant material can lead to a wide variety of toxic responses, which of course depend upon the chemical nature of the toxin involved. The responses can be reasonably mild, such as an upset stomach, or can be life threatening, such as in the case of cassava root. The effects can be direct and immediate or they can be indirect and cumulative over time. As an example, vegetables within the Brassica family (i.e., broccoli, brussels sprouts, cabbage, cauliflower, and others) contain considerable amounts of goitrogen, a chemical that interferes with the uptake of elemental iodine. Iodide deficiency causes the thyroid gland to enlarge, potentially resulting in a goiter.

Toxins in plants may have evolved for a variety of reasons, including pesticide resistance, as plants, over the course of evolutionary time, wage chemical warfare against infectious agents such as fungi as well as direct consumption by insects and other herbivorous animals, including humans. Since herbivory, the consumption of plant material, often involves the consumption of leaves or stems rather than the entire plant, a bad gastronomic experience may be a stern enough warning to ensure that future grazing is diverted toward more palatable plant material.

Unlike herbivores that often consume only a small portion of a plant while grazing, many carnivorous animals ingest their prey items whole; therefore the defensive poisons necessary to deter predation often lead to much more severe effects than a simple upset stomach. To be a real deterrent, animal poisons need to pack a wallop. For example, the poisons within the skin mucous of poison dart frogs or the flesh of a puffer fish are as likely to kill the consuming predator as they are merely to convince them to move on and make a different dining selection.

In nature, the art of chemical warfare may have reached its zenith with the innovation of venomous animals, those that not only contain poisonous toxins but also have the anatomical apparatus to inject those toxins directly into other animals. Venoms come in four different types: *cytotoxic*, causing cell death; *proteolytic*, dismantling the molecular structure around the area of the injection; *hemotoxic*, causing failure within the cardiovascular system; or *neurotoxic*, acting on the nervous system and the brain.

Venoms are used as defensive tools in some insects (wasps, ants, bees) and fish (lionfish), but can also be used as prey immobilizers for other animals such as fish-eating mollusks (conefish) as well as some spiders and snakes. Venoms are only rarely simple solutions of one or a few compounds, but rather are mixtures of chemicals that often have multiple impacts on the injected victim. Snake venom presents a classic portrait of an animal venom, and is highlighted below.

Snake Venom

For a snake, a limbless and relatively slow-moving animal, to prey upon a hyperactive rodent (rat or mouse) the tables have to be turned. For nonvenomous snakes, the capture of speedy, sometimes dangerous prey (rats and mice can bite!) is accomplished through stealth, followed by overwhelming strength. Constrictors do not use venom, but rather, after a bite, they quickly throw coils around the victim, constricting it to impair breathing and circulation. For a venomous snake, the act of predation is slightly different, as they quickly inject the prey animal with venom and then wait until death or immobilization occurs.

Snake venom is modified saliva that is stored in a specialized structure and has been augmented with a series of toxic proteins. Despite the wide array of proteins that may contribute to the toxicity of snake venom (no fewer than twenty-five), the impacts of venom fall loosely into two categories: venoms that impair the circulation of blood, and venoms that impair the electrical connections between locomotor nerves and muscles. The pit vipers of the Americas—the rattlesnakes, for example—generally employ venoms that target circulation, while the kraits of Asia and the mambas of Africa employ neurotoxic venom.

From a toxicological perspective, the bite of a rattlesnake, should it choose to give its victim a full dose of venom, is an ugly affair. In most cases, rattlesnake venom leads to blood loss due to the rupture of red blood cells and secondary bleeding. The loss of blood volume also leads to a lowered blood pressure and to shock. Along with the breakdown of red blood cells, proteins within the venom begin the process of tissue digestion. The wound site associated with a rattlesnake bite is black and blue, swollen; the tissue can be irreversibly damaged and deformed; and it can all be exceedingly painful.

In contrast, the bite from a true viper, which contains neurotoxin, is much cleaner—but for many of these snakes, more lethal. Neurotoxins have raised the bar of animal chemical combat to an entirely new level. To understand how this is true, it is important to understand a little bit about how nerves and muscles are able to propagate electrical signals.

Electrical Signals and Their Propagation: The Achilles Heel

Electrical signals are propagated in the brain and descend down the spinal cord. At that point, the signal moves from the spinal cord to the muscle via a nerve cell extension, the axon. The axon is an organic wire; that is, it is a cellular extension that allows for the propagation of electrical signals from the spinal cord to remote muscles that can be considerable distances from the spinal cord.

The electrical signal conveyed by the motor neurons (the nerve cells that send signals to the muscles) is as simple as Morse code. But rather than a collection of dots and dashes, it is merely one signal, effectively the "dot" of Morse code. Perhaps an even better metaphor is the binary language of computers, where the signal is either an "0" or a "1". The nerve signal is the result of the flow of ions, and is the same for virtually every animal (see chapter 2). To create the "dot" of an electrical signal, there has to be an upswing (the formation of the dot) followed by a downswing (or a return to baseline conditions). In electrical tissue, the signal is a change in polarity, or charge, of the cell membrane. At rest, the membrane of the nerve axon is slightly negative; during the upswing of the signal (also known as an *action potential*), positive charge flows across the membrane into the axon proper; and during the downswing of the signal positive charge flows back out of the axon.

The changes in ion flow during an action potential are the result of the opening and closing of ion channels or gates. At rest, the concentration of positively charged sodium ions outside of the axon is approximately twelve times greater than it is inside, whereas the concentration of positively charged potassium ions is approximately forty times greater inside the axon than outside. Consequently, during the upswing of the action potential, ion-specific gates open and sodium enters the axon. Likewise, during the downswing of the action potential, ion-specific gates open and potassium leaves the axon. The overall change in ions inside and outside the axon during a single action potential is actually quite small, but during a rapid activity event (like running a marathon, for example), the number of action potentials that are sent down the motor neurons innervating the muscles of the legs are so vast that the overall inflow of sodium and outpouring of potassium from the accumulated action potentials can begin to run down the ion gradients. To maintain a constant ionic difference inside to outside of the cell, an energy-requiring pump, the sodium-potassium ATPase pump, pumps sodium back out of the cell and pumps potassium back in.

At the nerve terminus, where the end bulb of the neuron interdigitates with a muscle fiber, there are a few more proteins involved in the cellular communication that leads to coordinated movement. The electrical signal—that is, the action potential that has run from the nerve cell body in the spinal cord all the way to the end terminus—changes slightly and allows for the inflow of calcium ions rather than an inflow of sodium ions. These ions, in turn, lead to the release of vesicles from golgi bodies found in the end terminus of the neuron. *Vesicles* are spheres of lipid bilayer surrounding an inner proteinaceous solution, much the same way that a jelly donut is a sphere of dough surrounding an inner solution of grape jelly.

Upon their release from the golgi apparatus, the vesicles migrate to the membrane of the end terminus, where they dump their contents into the open, fluid-filled space between the end terminus of the neuron and the membrane of the muscle fiber. For virtually all animals, the vesicles are filled with a protein, acetylcholine, which then migrates across the gap between nerve and muscle and binds to a receptor on the muscle membrane. When stimulated, the acetylcholine receptor opens a channel that allows sodium to flow across the membrane of the muscle, and the electrical "dot" that transcended the neuron can begin its journey across the muscle fiber, ultimately leading to a fiber contraction.

For the electrical signal to be completed, one final event is necessary: the turning off of the signal. Acteylcholine will bind the cholinesterase receptor reversibly, such that it will bind, activate, then release. However, if the acetylcholine molecule remains intact in the synapse, the space between neuron and muscle fiber, it can reactivate the receptor. To turn the signaling chemical off once and for all, yet another protein, acetylcholinesterase, is found embedded in the membrane of the muscle fiber. Acetylcholinsterase cleaves the molecule into acetate and choline, and puts an end to the activation of the acetylcholine receptor.

When looked at as a system, the action potentials that stimulate coordinated muscle function require no fewer than six proteins to act in a coordinated fashion. The sodium, potassium, and calcium gates, the sodium-potassium pump, the acetylcholine receptor, and acetylcholine esterase all have to work together for locomotion to occur. While exquisite in its performance and in its ubiquity (all animals basically share the same system), it is the Achilles heel of animal locomotion. Chemicals that alter the performance of any of these proteins lead not only to uncoordinated spasms and tremors, but ultimately to a rapid death through their action on muscles involved in breathing and on the muscle of blood circulation, the heart. Indeed, the more elegant the system, the easier it is to "clog the pipes."

Clogging the Pipes

As stated previously, snakes are slow ambush predators that prey upon fast, dangerous prey species. To even the odds, snakes with neurotoxic venom inject chemicals into the victim that jam all communication between the upstream nerve cell and the downstream muscle fiber. The injected rat or mouse loses motor control and can be easily dispatched and swallowed. Interestingly, different neurotoxic venoms from snakes attack different proteins or processes within the nerve–muscle junction. For example, one of the toxins found within the venom of the krait (α-bungarotoxin) jams the acetylcholine receptor permanently open while a second protein (β-bungarotoxin)—*from the same snake species!*—blocks the release of acetylcholine-laden vesicles from the golgi apparatus. The black mamba, in contrast, injects a venom that contains α-neurotoxin, which binds to the acetylcholine receptor, thereby preventing acetylcholine from successfully binding to its receptor.

Elevating chemical combat to greater levels is not just within the purview of snakes but has also been accomplished by other animals. For example, cone snails use stealth to capture their prey, fish, which are harpooned by their projectile proboscis and injected with conotoxin, which blocks calcium influx at the end terminus of nerve axons. And it's not just predators that have entered the chemical arms race. The poison arrow dart frogs from parts of tropical South America exude epibatidine onto their skin, which acts on predators to block the acetylcholine receptor, thereby breaking the connection between neuron and muscle fiber. Non-animals also get into the act, as the plant *Strobanthus* produces ouabain, a toxin that binds to the sodium-potassium pump, rendering it inactive. Furthermore, mushrooms such as *Anabella* produce the toxin curare, which acts to block the acetylcholine receptor. Bacteria also play the chemical warfare game and the botulinum neurotoxin A—the exotoxin produced by the bacteria *Clostridium botulinum*—is generally considered to be the most toxic chemical known. Any exposure to this neurotoxin, even at the nanogram-per-liter level, is fatal, as it prevents the upstream neuron from releasing acetylcholine.

The first poisons were natural, and while some of these toxins apply the blunt-force approach of cell trauma and death, others are rapier-like in their precision. The venoms that attack the exquisitely designed neural system can knock out one or more simple proteins and cause dramatic chemical mischief. While some organisms, under evolutionary pressure, developed the chemicals to derail nerve to muscle communications, humans have also done so—either deliberately or, as will be seen in subsequent chapters, by pure coincidence.

Chapter 9

Metals: Gift and Curse

The gods, who live on Mount Olympus, first fashioned a golden race of mortal men.
— Hesiod, *Works and Days*

The previous chapter, regarding animal poisons, focused on venoms and poisons that occur naturally and are deliberately synthesized by organisms. Most of the remaining chapters in this book focus on chemicals that are anthropogenic in nature and that cause environmental pollution as well as adverse health impacts. Pollutants can be naturally occurring compounds whose presence in the environment is altered by human activity, they can be chemicals synthesized or purified by humans from other sources, they can be by-products that form during such synthesis, or end products of the consumption of chemicals during human activities (e.g., the burning of fossil fuels).

With regard to the class of chemicals that humans purify or synthesize, it makes sense to begin with metals, for a variety of reasons. The first is historic, as metals were one of the first materials to be extracted and refined by early civilizations. A second is complexity. Metals, in the form most bioavailable to animals, are single atoms, and when found in water they are individual metal ions. As such, they are chemically the simplest of the toxic compounds, as they are only one single atom in size.

So what is it about metals that make them so historically important? Metals, as industrial materials, are malleable, capable of being stretched into a wire; ductile, capable of being hammered or compressed into a thin sheet; and good conductors of heat and electricity. Metals allowed early civilizations to develop a level of craftsmanship that was impossible when working with stone or wood. In contrast to wood, stone, or pottery, metallic items can be crafted into innumerable different shapes. Furthermore, the value of the metal never diminishes, for even if a vessel cracks or a sword tip is broken off, the metal can always be melted down and reformed. Indeed, the concept of turning swords into ploughshares was a literal reality long before it became an idiomatic expression of peace. Metallurgy changed ancient human history just as organic chemistry changed modern human history. The introduction of a pliable and reusable material eventually infiltrated virtually every aspect of life, and in many ways solved problems that had plagued humans for generations.*

Unfortunately, metals sought by humans for use in tool development also proved to be toxic when mishandled or misused; therefore the historical development of metal refinement came with a toxic cost—the increased incidence of metal ion poisoning. For example, consider the parallel histories of lead, wine, and concentrated sugars during the Roman Empire. Two thousand years ago, the wine preference among Roman citizens tended to be sweet and strong. Since concentrated sugars were rare during this time period, sweetening of

* The significance of metals in ancient civilizations is exemplified in *Works and Days*, an 800-line poem written by the Greek poet Hesiod over 2,700 years ago, in which he laid out the five Ages of Man, three of which were metals (gold, silver, iron) and a fourth that was a metal alloy (bronze). Furthermore, when the Roman poet Ovid (first century BCE – first century CE) reduced the Ages of Man from five to four, the age that he dropped, the Heroic Age, was the only nonmetallic one.

the wine was accomplished either by adding honey or by boiling the wine into a sweet, syrupy concoction known as *sapa*. Boiling wine into *sapa* was accomplished with the use of a flat, lead-lined vessel, and during the process, lead acetate, a sweet-tasting neurotoxin, was also produced. While the use of the lead-lined vessels helped to satisfy the demand for concentrated sugars, it also very clearly caused lead toxicity in those imbibing in the syrupy product.

Another, though less dramatic, example of the joint blessing and curse of refined metals is evident from the relationship that ancient civilizations had with *amurca*, the watery liquid that flows from olives when initially pressed under light pressure, and often concentrated by boiling in a copper vessel. The thickened mixture was then used for a variety of purposes, including the killing of insects and weeds. When *amurca* was boiled, copper compounds from the metal would form in the solution, enhancing the toxicity of the *amurca* and thereby enhancing its effectiveness as an insecticide and an herbicide.

The interplay between metal utility and toxicology continues to this day. In a modern example, the low melting point of lead, as well as its incredible plasticity, has made it valuable as a solder, capable of binding other metals together. Lead solder was used in the production of food cans and in household plumbing, despite the fact that, in both cases, lead ions could potentially leach from the solder, poisoning these sources of food and drinking water. Even though lead's toxicity has long been well known, legislation restricting its use in household water pipes was not passed in the United States until 1986, and the ban on lead solder on food cans did not come into effect until 1995. Metals, while incredibly useful, come with a toxic cost, one that has been carried throughout modern history.

Molecular Metals: A Matter of Ions

While the utility of metals, as a building material for tools, is based upon its metallurgic properties, its toxicity is based upon its elemental properties—the atomic structure of the metal ions themselves. The periodic table of the elements helps to put this atomic structure into context. Metals on the periodic table include the alkali metals in group 1 (the leftmost column), the alkaline earth metals in group 2, the transition metals in groups 3–12, and a stair-stepping group of seven "other metals," including tin, aluminum, and lead on the right-hand side of the table. The position of metals on the periodic table tells us about their ionic disposition. For example, the alkali metals have a simple ionic chemistry in that, when isolated in water, they all lose a negatively charged electron and become positive ions. Sodium and potassium ions therefore carry a +1 charge, are positive (monovalent) ions in water, and have an oxidation state of 1. Likewise, the alkaline earth metals, when isolated in water, lose two electrons and are divalent (or 2+) ions with an oxidation state of 2.

Within the middle portion of the periodic table are the transition metals, elements that differ from the others in that they are stable in more than one oxidation state. For example, copper ions can have an oxidation state of +2 or of +1, and they are equally stable.

The transition metals are of interest biologically in that they come in two forms: those that are essential and those that are not. The essential transition metals are vital for survival and growth and include copper, zinc, manganese, and iron, among others. These are the metals found in most multivitamins, and as such are micronutrients. Essential metals act as necessary chelation or binding agents for enzymes and cell membranes; for example, iron is an important component of hemoglobin, the primary oxygen-carrying protein within vertebrate red blood cells. While less well known relative to iron, zinc is also an essential micronutrient, as it is a necessary cofactor involved with more than 300 different proteins. On a cellular level, the acquisition of the essential metals is vital, and as such, specialized and efficient mechanisms have developed to acquire these essential metal ions from the environment.

Nonessential metals are found in the same columns of the periodic table as are the essential metals. For example, the essential micronutrient copper is found in the same column as the elements silver and gold, both of which are toxic. Likewise, the essential metal zinc is found in the same column of the periodic table as the toxic elements mercury and cadmium. These columns are also referred to as elemental groups or families, as they represent more than just a simple classification scheme for the elements. Elements within a family share a similar orientation of the electrons in their outermost electron shells, and these similarities give the elements similar chemical properties. Importantly, most biological transporters, the proteins that transport metal ions into or out of cells, cannot distinguish between elemental family members. Therefore, efforts put forth by cells to acquire essential metals, such as zinc, can also lead to the acquisition of nonessential and highly toxic ions, such as cadmium or mercury.

Absorptive Pathway for Metals

Relative to toxicity, the bottom line for many of the transition metals is the speciation of the metal into an ion or ion complex and the interaction of these ions with cell receptor sites. Since ionization occurs when materials are in water, absorption of metals into animals generally occurs through protein-mediated transport of ions from an aqueous environment (water or the aqueous slurry within the digestive tract) into the blood.* Consequently, the primary route of exposure to metals in terrestrial animals is generally via the ingestion of ion-containing food or water. For aquatic organisms, ingestion is augmented by absorption across the gills.

With respect to essential transitional metals specifically, it is important to recall that animals run the risk of both micronutrient deficiency and toxicity (see chapter 1). For terrestrial animals, the diet is the only source of essential micronutrients, and the amount of these micronutrients in food (regardless of whether the animal is an herbivore or carnivore) is quite low—which is, after all, why they are

* Hazardous occupations (e.g., working in smelting operations) or lifestyle choices (e.g., smoking) can result in the absorption of metals from the atmosphere through smoke or fumes, but beyond these unique routes of exposure, aqueous exposure predominates for most metals.

called micronutrients in the first place. As such, efficient-transport proteins have developed, across evolutionary time, to transport micronutrients that enter the digestive tract when food is consumed.

While the task of the carrier proteins in the intestine is fairly straightforward—transferring essential metals from the gut into the blood—the job is made much more difficult by the heterogeneous nature of the gut contents, as well as the variability within digestion. A number of factors influence the transport of essential metal ions from the gut, including the gastrointestinal contents, the chemical form and the environmental matrix in which the ingested metal is contained, and the individual's dietary habits and nutritional status.

Once a metal ion has worked its way free from the matrix of the gut contents, it can be transported into the animal. As stated previously, the transport relies upon metal-carrier proteins. On the gut content side of the cell epithelial layer, the free metal ion has to interact with one type of carrier protein in order to cross the lipid bilayer into the cell. On the blood side, the ion again has to pass from the cytoplasm of the cell, across the lipid bilayer, and into the blood. Interestingly, for most ions the proteins involved in this transcellular transport are different from one side of the cell to the other.

Storage of excess, but essential, metals can be accomplished via the use of a group of intracellular proteins known as metallothioneins. Metallothioneins are proteins rich in the amino acid cysteine, which allows for aggressive binding of the protein with positively charged metal ions. The importance of metallothioneins is evident from their pervasiveness. Animals, plants, and bacteria all contain small, cysteine-rich proteins that perform the functional role of reversible metal binding and storage. Furthermore, there is a very strong genetic consistency in the metallothionein protein structure across all vertebrate animals.

For many metal ions, particularly those within the same column of the periodic table, the need for essential metal ions runs counter to the need to minimize the uptake of toxic, nonessential metal ions. For example, the epithelial layers of the intestine have to remain efficient at the uptake of the essential metals copper and zinc, but this efficiency proves problematic relative to the uptake of the sister metals silver, gold, cadmium, and mercury. These sister elements (as well as some other transition metals) are all capable of being transported by the same transport protiens, allowing them to leave the gut contents and gain entry into the bloodstream of the animal. Likewise, the metallothioneins allow for the efficient conservation and storage of excess essential and nonessential metal ions alike.

While the origin of metallothioneins was probably as a mechanism for sequestering essential metals, metallothioneins are also useful in impeding the flow of nonessential ions toward intracellular target sites. In fact, if the number of binding sites for a nonessential metal ion is exceeded by the number of binding sites on all of the metallothioneins in the cell, the flow of ions into the cell will be effectively counterbalanced by the binding capacity of the metallothioneins, thereby impeding the toxic effect. It has been suggested that an increased capacity to withstand the adverse impacts of metals in animals and humans may be due to the excessive genetic expression of metallothionein protein. The protein can act as a sponge in the presence of toxic metal ions, locking up the ions in a bound and harmless form. Toxic effects would only occur in a spillover situation in which the number of metal ions in the cell overwhelmed the available metallothionein binding sites, thereby exceeding capacity and allowing the unbound metal ions to bind to biological receptor sites, with potentially disastrous, even fatal, results.

The absorption of essential metals from the intestinal contents into the blood is under feedback control, such that deficiencies in metals can enhance their uptake from the intestine. Unfortunately, this can have adverse consequences relative to metal uptake in individuals with dietary imbalances. For example, it has been reported that iron deficiency in malnourished children is associated with increased blood lead concentration. In animal models (e.g., rats) it has been shown that iron deficiency increases the gastrointestinal uptake of lead, perhaps due to an increase in the number of iron-transport proteins. In the absence of iron, the transport proteins may be transporting lead instead. A similar condition has been demonstrated for alcoholics, for as the disease advances, individuals acquire more and more of their daily calories from alcohol. A paucity of essential micronutrients in the diet, such as zinc, will increase the effectiveness for zinc uptake, despite the fact that the zinc ions are not available in the limited gut contents. Since alcoholics are often tobacco smokers, and since tobacco leaves contain considerable amounts of cadmium, the intestinal metal transporters can begin to transport the nonessential and toxic metal cadmium in lieu of zinc.

Metals and metal pollution pose a different type of risk to fish, as the fish gill does double duty, acting as both a respiratory and an osmoregulatory organ at the same time. A fish gill looks like a radiator, with baffles and extensions to pack as much surface area into the oral cavity of the mouth and throat as possible. Because of the large amount of surface area in a gill, and the small diffusion distances between the blood and the ventilated water, fish prove to be highly porous animals that are constantly transporting ions in one direction or another across their gills.

In seawater, the concentration of ions (generally ions from the alkali and alkaline earth metals, e.g., sodium, potassium, magnesium, and calcium) exceeds that in the blood of the fish, and these ions diffuse, through protein pores and channels, from the water into the animal. Despite the influx, the blood concentration of these ions has to be rigorously controlled for physiological reasons; therefore the influx of ions from the water into the blood is fatal if not regulated. Consequently, the fish is constantly and actively transporting the key alkali and alkaline earth metals from the blood back into the seawater. A freshwater fish, in contrast, is constantly losing these same ions to the water, where they will have to be actively reacquired through protein-mediated active transport.

In polluted seawater or freshwater, the concentration of nonessential transition metals in the water can be so high that their gill uptake can be fatal to fish. Given that the gill in a freshwater fish is designed to reacquire lost ions, the freshwater fish in a polluted environment is fighting two contradictory needs: the need to acquire essential ions such as sodium and potassium from the water and the need to prevent nonessential ions of transition metals from entering the animal. In the presence of toxic metals, as the examples below illustrate, the animal can fatally fail on both accounts.

Two of the important proteins involved in the transport of ions across the gills are the sodium-potassium transporter and the magnesium-calcium transporter. Called "pumps," these transporters consume cellular energy and require an atom of zinc to be bound to the protein to function properly. In polluted environments with an excess of copper or silver (note that these two elements are from the same family in the periodic table!) in the water, these ions will bind to the sodium-potassium transporter, kicking the zinc off of the binding site and rendering the pump useless.

In this case, a freshwater fish will desalinate (experience low sodium) and die. Likewise, in polluted environments with an excess of cadmium or mercury (again, note: same family on the periodic table!), these metals will bind with the calcium-magnesium transporter and the fish's blood will decalcify and again death will ensue.

If the fish survives the initial influx of transition metal ions, it may be able to undergo acclimatory responses that will reduce the gill uptake of transition metals. This can occur by decreasing the rates of metal influx into the gill cells. The fish can also alter the production of metallothioneins, thereby tying up the toxic metal ions before they can bind to the transporter proteins. The fish will live if the flow of the metal ions can be attenuated enough to allow the sodium-potassium and the calcium-magnesium pumps to continue to function. The fish dies if the metal influx overwhelms the cellular defenses, binds to the pumps, and reduces their performance.

In natural waters, the toxicity of metal ions comes with a caveat, as water quality dramatically influences toxicity. When considered in composite, the gill tissue can be considered to be a binding site, or *ligand*, for waterborne ions. (The reality, of course, is more complicated, as the gill is not a homogenous ligand; however, in this case it is useful to consider it as such.) In freshwater devoid of colloids, the fish gill may be the dominant ligand that is binding to the elemental ions that are dissolved in the water passing across the gills. Uptake of these ions will occur virtually unabated, as fast as the gill proteins will allow, as there may be no other ligands competing for them.

Toxic metal ions, such as cadmium or mercury, can also enter the fish via this same mechanism, active uptake across the gills, and the toxicity of these ions is often related to the composition of the particles carried in the water. In waters carrying other ligands (for example, colloidal material), the colloids can be considered to compete with the gill for the elemental ions. In much the same way that plasma proteins mediate the delivery of lipophilic steroids to target tissues within the body of an animal, humic acids and other water-soluble organics can decrease the free concentration of metals in water, thereby reducing the toxicity of the metal ions to the fish. This is the basic explanation behind the biotic ligand model, a conceptual model that suggests that the toxicity of metals in water is not uniform, but depends upon characteristics of water quality as well as the other compounds solubilized in the water.

For ancient human civilizations, the use of transition metals revolutionized tool development while also altering geochemical cycling of these normally rare metal ions. The altered cycles placed animals and humans alike in environments where the inhalation of vapors and metal ion ingestion upped the ante with respect to human toxicity. At the same time, the extraction processes caused environmental concerns that were a rarity prior to metal exploitation. In short, metal toxicity has been the devil in the details of human metal extraction, use, and tool development, and continues to be so to this day.

Chapter 10

Combustion

Workin' in a coal mine, Goin' down down down
Workin' in a coal mine, Whop! about to slip down
Five o'clock in the mornin', I'm already up and gone
Lord I am so tired, How long can this go on?
– Allen Toussaint, "Working in the Coalmine"

If a conversation regarding chemical pollution begins with metals, fuel combustion cannot be far behind. In fact, the two are inextricably linked, as the extraction of metals from ore, also known as smelting, requires considerable heat. Combustion, whether employed to coax metals from ore, to cook dinner, or simply to heat a home, is a messy business in its own right. Whether the fuel combusted is wood, coal, or oil, the consequences of combustion are soot, ash, and smoke. As we shall see, exposure to these end products comes with a significant toxicological price tag.

While the cultural history of metal use and extraction is, inevitably, closely linked to the industrial use of fire, the toxicological consequences of the two are completely different. The story of metal toxicity, as discussed in the previous chapter, involves cellular or molecular misidentification, in which toxic metals are mistaken for benign ones. In contrast, the story of combustion involves cellular sloppiness, in which the process of eliminating toxic compounds goes wrong, potentially leading to dramatic, adverse toxicological consequences.

Combustion

A major attribute of life is the process of *anabolism*, the construction of complex organic compounds, cellular components, cells, and tissues, from simpler compounds. Likewise, a major attribute of death is the process of metabolism, the breakdown of the complex chemicals of life back into the simpler compounds from which they were initially formed. If a living organism dies and decomposes quickly in the presence of ample oxygen (so that aerobic metabolism can occur), its organic material can be completely broken down into component parts that can be recycled into other living organisms. As long as the process remains aerobic, the end products can be as simple as carbon dioxide and water, which are then liberated back into the environment to be recycled. While many forms of life, such as animal tissue, decompose rapidly under aerobic conditions, other forms of organic matter, such as wood, are relatively resistant to decomposition. The woody remains from dead trees will endure to await combustion long after the tree has died.

Although wood, when compared to animal tissue, is resistant to decomposition, it ultimately *will* decompose, given enough time. However, under unusual circumstances the decomposition of wood and other organic material can be sidetracked. In this case, aerobic decomposition is hindered and the organic material does not break down completely, but rather the molecules are converted, often over geologically long periods of time, into other complex organic materials. These anaerobic processes ultimately lead to the production of fossil fuels such as natural gas, oil, oil shale, and coal.

Wood, natural gas, oil, and coal are all *fuels*, that is, materials that contain potential energy in forms that can be liberated to do useful work, and the process by which this liberation takes place is known as *combustion*. For example, the primary chemical in natural gas, methane, is the smallest of the hydrocarbons, as it consists only of one atom of carbon and four atoms of hydrogen. When completely combusted, it produces carbon dioxide and water. However, even the combustion of methane is not always complete, and alternate end products, such as carbon monoxide, can be produced as a consequence of inefficient or incomplete combustion. As fuels increase in their chemical complexity from natural gas to oil, coal, and wood, the chemical consequences of incomplete combustion are realized through the release of smoke, soot, and ash, all of which contain a suite of combustion by-products.

Toxic Consequences of Combustion: The Climbing Boys of England

The toxic impacts of combustion by-products were first documented in the 1700s in children, more specifically in the "climbing boys" of England. In this era, chimney sweeps, rather than being the lovable characters from *Mary Poppins*, were a group of hardscrabble individuals who made their living by working under incredibly hazardous conditions. The chimneys of London were irregularly shaped, and soot within them had to be cleaned out by hand. Adults were too large to scale the inner workings of the chimneys; therefore climbing boys, very often orphans as young as four years old, were employed as "apprentices" to shimmy into the recesses. Working conditions for these children could not have been worse, with soot raining down on them from above as they cleaned. The heat, air quality, and tightness of the space created potentially lethal conditions. Due to the narrow passageways and the insufferable heat, many children climbed the passageways naked, exiting the chimney sweaty and covered in soot.

In 1775, an English medical doctor and surgeon, Dr. Percivall Pott, took a closer look at the adult men who had once been climbing boys. During their work, the sweat would run down their bodies, depositing soot in the groin. These boys had horrible hygiene, with some having only a single "annual wash." Consequently, soot containing carcinogenic material would imbed within the skin of the scrotum and remain there for long periods of time. These exposures resulted in "soot-wart," a ragged and painful sore on the scrotum with hardened edges. Over time the associated cancer would spread to the testicles and into the abdomen, with devastating results. While it would take another century to elucidate the chemical responsible for this cancer, Dr. Pott had documented the first example of an occupational hazard leading to tumor formation.

It's interesting to note that scrotal cancer caused by chimney soot was completely preventable. The cancer was not prevalent in chimney sweeps working on the European continent, but rather appeared limited to the climbing boys of England. On the continent, the chimney sweeps wore protective clothing and also had much better hygiene, bathing more frequently. Once these two practices were adopted in England, the cancers effectively disappeared.

Toxic Consequences of Combustion: Polycyclic Aromatic Hydrocarbons

Despite the results from the "climbing boys" study, the list of known chemical carcinogens grew slowly. Prior to 1950, only three chemical mixtures were known to cause cancer in humans: coal soot, tobacco, and naphthylamines from the production of synthetic dyes. Tobacco, like coal tar, had been recognized as a carcinogen ever since another London physician working in the 1700s, Dr. John Hill, first observed that excessive use of smokeless tobacco, or snuff, led to nasal cancer.

Unlike coal tar and tobacco, the link between naphthylamines and carcinogenicity was not discovered until the 1870s. Prior to that time, all industrial dyes were derived from natural sources. However, with the advent of synthetic fuchsine dyes in Germany came reported cases of bladder cancer. When World War I interrupted the industrial cooperation between Germany and the United States, an American firm, the DuPont Company, began to synthesize dyes. The specter of bladder cancer as an occupational hazard for those who worked with synthetic dyes also crossed the Atlantic, leading to the disturbing conclusion in 1947 from Dr. Arthur Mengellsdorf of Calco Chemical Company that "100 percent of the beta-naphthylamine workers at Chambers Works have developed bladder cancer."

While early epidemiological studies pointed a finger at these three chemical carcinogens, it took over a century before experiments performed on lab animals showed direct carcinogenicity. In 1915, two Japanese researchers, Katsusaburo Yamigiwa and Koichi Ichikawa, induced skin cancer on the ears of rabbits by painting them with coal tar a few times a week over the course of a year. Following their pioneering work, chemical analysis of coal tars and coal tar extracts began to isolate the carcinogenic molecules found within. In 1930, the chemical dibenz[*a,h*]anthracene was demonstrated to be carcinogenic, and soon thereafter carcinogenicity was also demonstrated for a similar chemical, benzo[*a*]pyrene. By 1940, the epidemiological data was finally supported by studies on laboratory animals that demonstrated a direct link between specific chemicals and carcinogenicity. And the result: chemical carcinogenesis was linked to two classes of compounds: the aromatic amines (in the case of industrial dye workers) and polycyclic aromatic hydrocarbons. Since that time, many other compounds have been found to be carcinogenic; however, it was the PAHs and the aromatic amines that first established the link between chemical exposure and cancer.

Polycyclic aromatic hydrocarbons (PAHs) are cyclic rings of carbon and hydrogen, with the smallest PAHs containing no fewer, and the largest containing substantially more, than three rings. The naphthylamines differ from PAHs in that they only contain two rings, and also contain nitrogen, in addition to carbon and hydrogen. PAHs occur naturally in fossil fuels such as oil, coal, and tar deposits, as well as some edible oils. They are produced by the incomplete combustion of organic material such as coal and oil, as well as during the combustion of biomass, such as wood and tobacco leaves. The production of PAHs during tobacco smoking is part of what gives the practice such a carcinogenic punch.

Cancer and PAHs

To understand the link between PAHs, or aromatic amines and cancer, it is valuable to understand a bit more about what cancer is and how it originates. *Cancer* is the formation of neoplasms: a collection of cells that are undergoing relatively autonomous growth. For virtually all cell types except neurons, a cell lives its life conducting the job of the tissue it inhabits (liver cells doing liver business, intestinal cells conducting the work of digestion, etc.). During this routine, the cell will periodically divide into two daughter cells via the process of mitosis. A cell's life is the continuing grind of grow, work, divide, grow, work, divide—otherwise known as the *cell cycle*. The cell cycle, as mundane as it may appear from the perspective of a single cell, is a highly coordinated activity that is regulated via intrinsic chemical signals. Should a cell remove itself from the drumbeat of life and become damaged or dysfunctional, it will inevitably be genetically called upon to induce an orderly cell death, known as *apoptosis*. However, should cells "go rogue" and no longer adhere to the cues for self-destruction, division can become autonomic and comparatively unorganized, with one possible result being the development of neoplasms, or tumors.

One mechanism responsible for the development of neoplasms is genetic mutation. *Genetic mutation* is the chemical or physical alteration in the structure of DNA that results in inaccurate replication of that region of the genome. When a chemical carcinogen covalently bonds with the DNA, it causes a *DNA adduct*. A covalent bond is one in which two atoms share a pair of electrons. Covalent bonds are strong and the only way to remove a DNA adduct from the genetic material is to excise it, effectively cut it out of the genome, including the genetic material to which it is attached. Interestingly, these adducts, once excised from the DNA, are ultimately excreted in the urine. Furthermore, the concentration of DNA adducts in urine can actually be measured as a risk factor for the development of certain types of cancers.

However, all DNA adducts are not created equal. Some are excised from the DNA and are effectively repaired. When others are excised from the DNA, though, an error is made during the repair, resulting in a mutation, or an alteration in the nucleotide sequence (the coding region of the DNA). Some of these dysrepair mutations can be found in regions of the genome that are not particularly important with respect to carcinogenicity, and, as such, do not lead to neoplasms. Others, however, are found in important areas of the genome—for certain areas are more likely to cause cancer than others—and can initiate the development of tumors. These susceptible regions include areas that contain proto-oncogenes and tumor suppressor genes. *Proto-oncogenes* are highly conserved genes encoding proteins that stimulate the progression of cells through the cell cycle. In contrast, *tumor suppressor genes* encode proteins that inhibit the progression of cells through the cell division cycle. As such, proto-oncogenes provide a gas pedal for cell division (and consequently a gas pedal toward autonomous cell division). Tumor suppressor genes provide a brake to control cell division. Mutation within either region can lead to uncontrolled cell division or autonomous cell growth—fertile ground for the development of neoplasms. This accumulation of genetic damage in the form of mutant proto-oncogenes and mutant tumor suppressor genes is one of the driving forces that convert an orderly cell cycle to uncontrolled cell division, or tumor formation.

The extent to which DNA adducts are formed in response to chemical carcinogens depends upon the chemical reactivity of the metabolite. Surprisingly, many parent polycyclic aromatic hydrocarbons are not particularly carcinogenic. With that said, the devil of carcinogenicity lies within the details of excretion, through metabolism, of PAHs.

Polycyclic aromatic hydrocarbons are lipophilic, and because of that they tend to aggregate in living organisms or sediment and are less likely to occur in water. As was discussed in chapter 5, organisms can metabolize lipophilic compounds such as PAHs via Phase I oxidation/reduction reactions, as well as Phase II conjugation. Phase I metabolism only slightly increases water solubility, and therefore only slightly increases rates of excretion for most PAHs. In contrast, Phase II conjugation increases water solubility much more dramatically. Unlike Phase II metabolism, which is fairly orderly and predictable, Phase I metabolism is a variable process with a suite of metabolites being produced during the process. Therefore, the profile for any PAH within the bloodstream of an exposed animal contains not just the parent compound but also an array of Phase I and Phase II metabolites.

PAHs and aromatic amines are carcinogenic and are rapidly converted in the liver into a suite of metabolites. Considering that the point of metabolism is to alter the solubility of the compound such that it can be excreted, it stands to reason that detoxification would not only decrease the lipid solubility of the PAH but also decrease its reactivity, or in other words, its propensity to form DNA adducts. However sensible that may seem, the converse is actually true. Highly reactive metabolites, such as epoxides and quinones, are often formed during Phase I metabolism. These chemicals actually increase, rather than decrease, chemical reactivity. Consequently, as many PAHs are metabolized and prepared for excretion, their carcinogenicity may actually increase.

It may seem disheartening that our own bodies are working against us with respect to chemical carcinogenesis. The flexibility of Phase I metabolism may be advantageous when an organism is confronted with a heretofore unknown lipophilic substance that needs to be converted into a water-soluble form in order to be excreted. However, at the same time, the level of flexibility within the system contributes to its capacity to produce carcinogenic metabolic intermediates. While this may be discouraging, it also holds out rays of hope. If the pathway of PAHs through metabolism can be redirected away from the toxic intermediates and toward the relatively benign water-soluble metabolites associated with Phase II metabolism, the likelihood of developing cancers can be substantially reduced.

In Greek mythology, Prometheus was a trickster who gave the gift of fire to humans—and he was punished severely for doing so. As history has taught us, we have also paid a price for the gift of fire: our long-time cultural intimacy with soot, tar, and ash, as well as the PAHs hidden within. We collectively pay with our health, in the form of an increased propensity toward chemically induced cancers.

Chapter 11

Drugs and the Toxicology of Addiction

Better Things for Better Living . . . Through Chemistry
— DuPont advertising slogan, adopted 1935

Drugs. The word conjures up a flurry of images. Pharmaceutical drugs have increased human life span and enhanced the quality of life for billions of people around the planet. At the same time, narcotics fuel illegal activity, wreak havoc on individuals, and devastate communities. Some drugs, such as alcohol and tobacco, oscillate in their favor with the general public, while others, such as coca, serve a relatively benign purpose when administered at low doses, but have a much more sinister character when taken at higher doses. This chapter will consider the toxicity of deliberately ingested materials that are taken to relieve pain, fend off disease, and alter the psychological state.

Self-Medication

Drugs are chemicals that, when absorbed into a person or animal, cause changes in cellular function. While a few drugs can trace their origins back to chemical synthesis in a laboratory, the vast majority of drugs have more natural origins. In fact, the first application of therapeutic drugs to alleviate pain or disease was through plant consumption. The practice of eating plants or other naturally occurring material for the therapeutic chemicals contained therein is known as self-medication, and the practice predates humans. African great apes chew on the pith of some plants, swallow leaves whole, and consume mixtures of dirt and the leaves of medicinal plants to fend off intestinal parasites and relieve gastrointestinal distress. Baboons and colobus monkeys also appear to engage in self-medication, primarily to minimize intestinal parasitism. Howler monkeys have been found to consume plants containing chemicals that are reproductively active, presumably to alter birth spacing and the sex of their offspring.

It seems likely that self-medication traveled the evolutionary path from nonhuman hominids to humans in a fairly direct and unbroken manner. One of our close relatives, the Neanderthal, was practicing self-medication with botanical material 24,000–30,000 years ago. Evidence for this comes from the dental calculus on the teeth of a long-deceased young adult Neanderthal. The buildup on the individual's teeth contains remnants of bitter-tasting (and nonnutritious) plant materials known to act as appetite suppressants. The possibility of a self-medicating Neanderthal is intriguing, as it suggests a level of sophistication with respect to botanical knowledge not previously credited to our long-distant cousins.

The transition from self-medication with plant material to the manufacture of compounds for use as medicinal drugs, i.e., pharmaceuticals, dates back at least 2,000 years. A few highly significant artifacts help to elucidate this development. For example, in 1989 a "first-aid kit" was pulled from the wreckage of 2,100-year-old Roman-era ship; the carefully sealed wooden chest contained pills made of ground-up vegetables, herbs, and other food plants such as celery, onions, and carrots. All of the pills (which have just recently been analyzed chemically) contain ingredients referred to in classic medical texts from the same era. These pills are the oldest pharmaceuticals that have been found to date, and their relatively simple composition (ground-up plant material) appears to be a stepping-stone between self-medication with biologic material and the pharmacological application of a medicine. A second artifact dating from roughly the same era is the Ebers papyrus, written around 1500 BCE. The papyrus, one of the earliest-known records relating to Egyptian medicine, is a scroll describing magical formulas and remedies; physical disorders such as burns, tumors, and parasites; and even mental disorders such as depression and dementia.

Psychoactive Drugs as Toxic Substances

The advent of pharmaceuticals may well be the first instance of "better living through chemistry," as drugs have historically reduced human pain and suffering. The history of drug toxicity is multifaceted, with many concordant story lines. Drugs that become toxic when misused, such as acetaminophen (chapter 5), the active ingredient in Tylenol, is one such story line, while drugs that become risk factors after chronic usage, such as tamoxiphen (chapter 14), is still another. But perhaps the best starting point for a discussion on drugs as toxic substances lies with psychoactive drugs and the toxicology of drug abuse.

First, consider for a moment the life of the average individual throughout ancient history. Exposure to the elements and backbreaking physical labor were the norm for many inhabitants of early civilizations, and day-to-day life must have been physically and mentally exhausting. It is not surprising that fatigue resistance would have been a highly sought-after attribute, and any natural product that provided it would have been highly prized. While the first cases of self-medication, involving the chewing of coca leaves or areca-betel, have been lost in the undocumented fog of prehistory, the therapeutic use of coca and areca-betel, in more or less natural and relatively low-potency form, for fatigue resistance has continued unabated up to the present day.

In addition to pain relief, psychoactive drugs were important in the lives of citizens living in ancient civilizations in at least two other important ways: relaxation in the form of "social time-outs" and the spiritual attainment of a different mental state. Alcohol may have traditionally been used much as it is today, being a consistent feature at festivals, religious ceremonies, and family rituals. These events may have given prehistoric communities a clear-cut time when acute intoxication was tolerated. Even with the occasional festival binge, drug abuse in ancient civilizations was probably rare due to the relative scarcity of alcohol and other psychoactive drugs, the social taboos associated with inappropriate use, and the strenuous nature of day-to-day existence. For most people, the luxury of chemical over-indulgence was probably too socially and physically expensive to be anything beyond an annual or semi-annual event.

While many of the first psychoactive drugs were probably pain relievers, as well as appetite, thirst, and fatigue suppressants, a third, more spiritual, class of psychoactive drugs was also in use at the same time. These drugs, known as entheogens or psychedelics, are hallucinogenic and have the capacity to induce altered feelings, thoughts, and perceptions of reality; they include the naturally occurring mescaline-containing peyote cactus and the psilocybin-containing mushrooms. While not widely used by the rank-and-file citizens of ancient cultures, they were in widespread use by their spiritual leaders.

Drugs, as much as any other class of chemical compounds, evoke Paracelsus and his adage that the dose does indeed make the poison. From the rudimentary practices of ancient Egyptians to modern-day pharmacology, any increase in the sophistication of psychoactive-drug processing leads to an inevitable increase in drug potency. Even during the earliest times of self-medication, people altered the chemical nature of drugs to enhance their delivery. For example, coca chewers increase the potency of coca leaves by adding an alkaline agent, such as lime, bicarbonate, or wood ash to the collection of leaves that they pack between their cheek and gum. While the potency of the alkaloids being released from the coca leaves are more diverse and are in much lower concentration than those found in cocaine, the chemical conversion that occurs when an alkaline agent is added to the leaves does allow the chemicals to be more readily absorbed across the inner skin of the cheeks, mouth, and oral cavity.

While enhanced potency increases drug efficacy, it can also increase the possibility of adverse toxicological outcomes, including death. A grim reminder of this potential lethality is the fact that drug overdose is the second-most common method used by women attempting suicide.* While any drug is dangerous when taken in excess, psychoactive drugs may be more so. For some, but not all, psychoactive drugs, exposure may trigger drug abuse that can lead to dependency and addiction.

Drug Addiction

Drug addiction generally follows a patterned response that begins with a psychoactive effect leading to compulsive drug-seeking behavior. It is the compulsive behavior of the user that characterizes drug addiction. The psychoactive effect of a drug occurs due to stimulation of nerve synapses in one of the pleasure, or reward, centers within the brain, very often involving the neurotransmitters serotonin, norepinephrine, and dopamine. These reward centers are extremely powerful, as early experiments with electrical stimulation have demonstrated. In studies where rats could press a lever and electrically stimulate their limbic system within the brain, the rats would press the lever almost continuously, up to 5,000 times per hour, at the expense of all other behaviors except for sleep. In humans, stimulation of one of the reward centers within the brain leads to feelings consistent with sexual orgasm, whereas stimulation of other regions leads to an overall feeling of well-being that overcomes negative thoughts.

With respect to these reward centers, it has been suggested that all roads lead to the neurotransmitter dopamine. Stimulation of two specific reward centers in the brain, both featuring dopamine as the

* For both sexes, self-inflicted gunshots are the number-one suicide method. In males, strangulation by hanging is number two.

primary neurotransmitter, has been implicated in many pleasurable behaviors as well as a number of different addictive behaviors, both chemically induced (e.g., addiction to alcohol, cocaine, and tobacco, to name a few) and non-chemically induced (e.g., addiction to gambling).

Psychoactive drugs that stimulate reward centers in the brain are extremely powerful and, in many cases, lead the individual down the pathway toward physiological addiction. And just as Paracelsus would have predicted, as the number of psychoactive molecules binding to neuronal receptors increases, so does the narcotic effect. This can be accomplished in two ways, either by increasing the number of neurotransmitters being released into the nerve synapse, or by decreasing the rate at which the neurotransmitters are removed from the synapse (by reuptake or metabolism). For instance, methamphetamine stimulates the release of the neurotransmitters norepinephrine, dopamine, and serotonin, whereas cocaine causes a reduction in dopamine reuptake, so that the dopamine is maintained in the neural synapse for longer time periods. Both processes cause narcotic effects by increasing the effective number of neurotransmitters in the nerve synapse, that tiny gap between adjacent neurons.

Beyond the "big three" neurotransmitters—norepinephrine, dopamine, and serotonin—other neurotransmitter receptors in the brain are also targets of exogenous psychoactive drugs. For example, heroin (morphine), perhaps the most addictive of the psychoactive drugs, binds to opioid receptors that are normally stimulated by a class of naturally occurring neurotransmitters known as endorphins, with the most common being β-endorphin. In a similar fashion, THC (tetrahydrocannabinol), the exogenous cannabinol found in marijuana, stimulates cannabinoid receptors that ordinarily bind to endogenous cannabinoids.

The three most popular psychoactive drugs—nicotine, caffeine, and alcohol—deliver their effects according to processes that are consistent with those discussed above. Nicotine, for example, binds to nicotinic receptors in the brain, stimulating an increase in the release of dopamine. Caffeine binds to adenosine receptors, indirectly resulting in an increase in glutamate and dopamine receptor efficiency.

In contrast with many psychoactive drugs that target a specific receptor, alcohol elicits its effect when an ethanol molecule binds to multiple neuronal receptors. In a similar manner as ether and chloroform, alcohol alters glutamate receptor function, resulting in muscular relaxation, loss of coordination, slurred speech, staggering, memory disruption, and in the extreme case, blackouts. Alcohol also enhances GABA receptor function, which induces feelings of calm, the reduction of anxiety, and sleep. (Valium has a similar effect on the GABA system.) Finally, alcohol also increases dopamine levels, similar to the actions of cocaine and amphetamine, thereby stimulating that very powerful reward center. Furthermore, alcohol increases levels of endorphin, leading to an endorphin "high" in a manner similar to that of morphine.

Drug Dependency

While some drugs can cause both addiction and dependency, *drug dependency* is defined by a two-way interaction between the drug and its neural receptor. Drugs that elicit drug dependency alter the cellular status quo, changing the number of neurotransmitters that the neural receptors respond to. The response of a neuron to an exogenous opioid illustrates the process. If an individual is using an opioid pain reliever, pain relief is based upon increasing the number of signaling molecules in the synapse between neurons. In response to the increased stimulation, the target neuron will attempt to reestablish homeostasis by reducing the number of receptors available to receive the signal. In the case of heroin, for example, the downstream (or postsynaptic) neuron can respond by reducing the number of opioid receptors embedded in the cell membrane. Once the number of receptors has been adjusted, the previous dose of the compound will no longer elicit the same effect, and the person being administered the drug will develop tolerance toward its psychoactive effects. The dose that once caused a euphoric high will no longer elicit the same response, and consequently no longer produces the same psychoactive effect.

People who have built up a drug tolerance often respond to the attenuated effect by increasing the dose administered. As the dose increases, the neurons readjust again by further down-regulating receptors. Consequently, addicts may not experience a euphoric, psychoactive sensation, but will continue using the drug to avoid uncomfortable feelings associated with a low level of neurotransmitter–receptor interaction.

When a drug user goes "cold turkey," the nerve cells do not immediately return to baseline conditions, because the number of receptors has been altered. During this lag time, the user experiences withdrawal. For some drugs (heroin, cocaine, methamphetamine, nicotine, and alcohol) the withdrawal period (the time during which homeostasis is reestablished) is painfully long, for it takes some time for the receptor molecules on the target neurons to be resynthesized. For other drugs (caffeine) the withdrawal period is much shorter because wholesale changes in the number of receptors do not occur as use of caffeine increases.

The distinction between drug addiction and dependency is important, as not all drugs that cause chemical dependency are addictive, and not all addictions are based upon substances that create chemical dependency. The best examples of agents that lead to addiction but not dependency are pleasure-generating activities such as gambling, which can become addictive when gamblers experience such strong cravings to continue gambling that they can lose control over their gambling, continuing to gamble despite its effects on their economic and family lives.

Interestingly, the natural hallucinogens, such as peyote and mescaline (and nearly all other hallucinogens) do not engender compulsive drug seeking or addiction, nor do they lead to chemical dependence. Part of the reason is that these compounds do not stimulate the dopaminergic reward center in the brain, as they lack affinity for the dopamine receptor and the dopamine uptake transporter.

Drugs as Exogenous Cell Signals

The psychoactive drugs discussed above elicit their action in many cases by acting as cellular signals. The drugs that bind to neuronal receptors are actually masquerading as endogenous neurotransmitters and are sending signals to the postsynaptic neuron, leading to a highly rewarding (though possibly devastating) sensation. Neurotransmitters are one important class of cell-signaling molecule, but there are others, including hormones. Hormones travel in the blood from the parent endocrine cell until they bind with receptors, either inside or on the membrane of receiving cells.

Interestingly, anabolic steroids (taken for their proclivity for building muscle mass and enhancing athletic performance) share features with psychoactive chemicals, despite the fact that euphoria is not usually the intended goal of steroid drug use. There is evidence from rodent studies that anabolic steroids are reinforcing drugs, meaning that the use of the steroids produces a psychological reward. This can be witnessed from conditioned place-preference studies, in which rats are administered an anabolic steroid in one of two adjoining chambers. Once chronically exposed to the steroid, the rats have been shown to exhibit a mild (though real) preference for the chamber in which the steroid was administered. Furthermore, other rodent studies have shown that anabolic steroids stimulate the dopaminergic reward center within the brain, although the mechanism does not appear to be stimulatory, in the fashion of amphetamines and opioids, nor does it appear to inhibit dopamine reuptake, as with cocaine.

Anabolic steroids are not only reinforcing drugs, as are many narcotics, but they also interfere with the normal function of the endocrine system, and as such are endocrine disruptors. The presence of exogenous anabolic steroids in the blood causes a negative feedback in the brain, specifically the hypothalamus and the pituitary. This feedback leads to a reduction in the synthesis of gonadotropins —water-soluble hormones that travel in the blood to the testes, where they stimulate the synthesis of testosterone. This reduction in testosterone in turn leads to a reduced capacity of the testes to produce viable sperm, a condition that may reverse only very slowly (in excess of two years) after the cessation of steroid use.

As with venoms and other naturally occurring poisons, psychoactive drugs and anabolic steroids high-jack the naturally occurring cell-to-cell communication systems. In the case of animal venom, the high-jacking is sinister and can be lethal. In the case of drug ingestion, the high-jacking can be therapeutic. From botanical self-medication to modern pharmacology, the therapeutic use of exogenous chemicals appears to be as old as our species itself. Response to psychoactive drugs and steroids is both psychological and physiological, but the benefit comes with serious potential costs, from dependency to addiction.

Chapter 12

70,000 Years of Pesticides

Great fleas have little fleas upon their backs to bite 'em,
And little fleas have lesser fleas, and so ad infinitum.
And the great fleas themselves, in turn, have greater fleas to go on,
While these again have greater still, and greater still, and so on.
— Augustus De Morgan

All who wish the Ulu well should daily repeat this motto: Do good carefully.
— Tom Harrisson

Like metals and drugs, a third functional group of chemicals—pesticides—has a long-standing role in human history. The first recorded remediation of pests by humans occurred more than 70,000 years ago, predating metallurgy, the domestication of animals, and the cultivation of crops.

In a cave in South Africa, layers of human bedding, primarily reeds and sedges, have been identified. Over time, the bedding material would have become infested with bedbugs and lice. To deter the bugs, the top lining of the bedding was layered with leaves from the tree *Cryptocarya*. For when these leaves are crushed, they release aromatic α-pyrones, compounds that repel insects. The architecture of the bedding layers suggests periodical burning, indicating that when the bug infestation got bad enough, the sedges were burned and new ones put in their place. While this approach toward pest mitigation may be humble, it is also abundantly clear that the chemical deterrence of pests, like the extraction of metals and the synthesis of rudimentary drugs, is a time-honored human tradition.

Further evidence supporting the use of insecticides dates back to the Sumerians and the Chinese between 3,200 to 4,500 years ago. As with the leaves lining prehistoric beds, the original pesticides used by these civilizations were readily available animal, plant, and mineral compounds. Mercury, arsenic, and sulfur compounds were all used for insect control in these ancient civilizations. Generations later, approximately 2,000 years ago, a botanical pesticide, dried chrysanthemum flowers, was added to the list of ancient pesticides.

Records from Greek authors during the fifth century BCE, as well as from Roman authors from the first century CE, suggest the use of a wide variety of plant extracts to control pests, including hemlock, bitter lupin, and absinthe, among others. Other common items used to deter pests included *amurca*, whose metal-associated history was discussed in chapter 9, as well as bitumen (solid or semi-solid petroleum), arsenic, wine (probably effective due to the alcohol it contained), burnt ashes (containing lye), seawater (and salt in general), and even stale urine (containing ammonia).

First-Generation Pesticides

The pesticide arsenal remained true to its humble origins until the twentieth century. Occasionally, botanicals or other mineral complexes would arrive on the agricultural scene and enter into use as pesticides; however for the most part, their arrival was more by happenstance than by any specific, concerted effort. As an example, tobacco, a botanical from North America, was introduced to Europe in 1559, and by the late seventeenth century Europeans who had colonized North America were using the leaves as an insecticide.

Prior to the era of second-generation pesticides, which began roughly in 1874 with the synthesis of DDT and reached full steam in the late 1940s after the end of World War II, pest management used both biological and chemical controls—with mixed results. While *chemical agents* were compounds applied to repel pests, *biological control agents* were organisms, and often were the pest's predators. For example, in 1888 the introduction of the Australian vedalia beetle proved to be a highly effective control agent against the cottony cushiony scale, *Icerya purchasi*. The adult insect attaches itself to a plant, become sedentary, and feeds on sap. The introduction of vedalia beetles was a godsend for the California citrus crop, and the effectiveness of the control continues to this day. At the same time that vedalia beetles were successfully controlling *Icerya*, biological efforts to control the gypsy moth and the boll weevil were failing for biological as well as cultural and political reasons.

The lack of across-the-board success of biologicals as pest-control agents was matched by equally unsuccessful efforts to control pests with chemicals. For example, in the late 1870s a group of arsenic-based pesticides, including Paris Green, London Purple, and white arsenic were tested by a group of Vermont entomologists who concluded that "no benefit was derived from the application of the poison." Years later, in 1915, arsenic returned victoriously to the pest-control scene, as the dusting of cotton crops with calcium arsenate proved to be a highly effective protection against the boll weevil.

Second-Generation Pesticides

Ultimately, with the dawn of second-generation pesticides, the tide turned away from biological control agents toward the chemical control of insect pests. Interestingly, it was not an agricultural crop, but rather a public health discovery that led the way. Sir Ronald Ross, a British doctor who allowed an *Anopheles* mosquito to feed on a malaria patient, observed the malaria parasite in the gut of the mosquito. His groundbreaking work demonstrated that mosquitos were the vector for malaria, and this discovery ultimately led to his receiving the Nobel Prize for Physiology and Medicine in 1902.

At the time when Ross was awarded his Nobel Prize, dichlorodiphenyltrichloroethane (DDT), the compound that would prove to be a champion mosquito-control agent, had already been synthesized. The compound's value as a pesticide, however, would have to wait for anther thirty-seven years, until 1939, to be discovered. When that notoriety came, it proved to be worth the wait. DDT was used to control malaria and typhus during World War II, and it proved to be such an effective pesticide that it won Swedish chemist Paul Hermann Muller, who uncovered the drug's hidden pesticide talents, the Nobel Prize in Physiology and Medicine in 1948.

After Muller's discovery, the American pesticide industry took to DDT almost overnight. The commercial production of DDT, which was negligible in 1943, increased dramatically, as over 2 million pounds of DDT *per month* were synthesized just two years later. Furthermore, DDT use spread to many different sectors, from mosquito control in rural and urban areas, to agricultural use on row crops and fruit trees, to the use in the gypsy moth–control programs in the northeastern United States, to use in livestock operations and other industries associated with food production. DDT was an overnight sensation.

Development of DDT and many of the other second-generation pesticides was enormously beneficial for a number of reasons. First, some of the first-generation pesticides, including mercury, arsenic, and lead, were abandoned for being too overtly toxic. In contrast, chemicals such as DDT were considered fairly benign. It is not uncommon to find photographs in which DDT was being sprayed to control mosquitos on beaches while children cavort in the mist. Second, some of the first-generation botanicals, such as dried chrysanthemum flowers, were difficult to obtain in sufficient quantities. Still others were either ineffective or broke down much too quickly to give long-lasting protection. In contrast, many of the second-generation compounds were oily and tended to stick to the plants, seeds, or household walls (in the case of DDT as mosquito control) to which they were applied. Furthermore, because many of these compounds also contained elements from the halogen column of the periodic table (which includes fluorine, chlorine, and bromine), they tended not to break down readily, and they remained in active form where they were sprayed. Finally, rather than having to be excavated or grown (again, as in the case of chrysanthemum flowers), the second-generation compounds were synthesized in a laboratory from fairly inexpensive materials. The ease of synthesis ensured that supply lines were no longer a problem, as these compounds could be produced in very large quantities.

Unfortunately, the very same properties that led to the popularity of second-generation pesticides were also their Achilles heel, and the cause of many of their environmental difficulties. The oily (in other words, lipophilic) nature of the compound, while allowing it to stick to leaves and walls, also allowed it to stick to skin, where it would eventually diffuse without the help of protein carriers into the blood of animals. The lipophilic nature of DDT increased the chemical's efficacy as a pesticide, as it allowed the chemical to rapidly enter insects, where it has its toxic impacts. Indeed, the lipophilic nature of these compounds also increases their capacity to absorb into (and decreases their capacity to be excreted from) non-target species, including humans.

DDT is not only lipophilic but it also shares a common characteristic with other halogenated compounds; that is, it persists in the environment. The *halogenated* organic chemicals contain one or more atoms of elements from the halogen column of the periodic table, which includes the elements: fluorine, chlorine, bromine, iodine, and astatine. DDT is a well-endowed halogenated compound, as it contains five atoms of chlorine. Halogenation allows it to persist in the environment, as it not only prevents it from breaking down once applied, but also prevents it from breaking down as it moves through the environment into non-target organisms. Consequently, the same properties that enhanced the efficacy of these compounds as pesticides also made them particularly dangerous when released into the environment.

Insecticides such as DDT elicit their lethal effects on insects in much the same way as do naturally occurring neurotoxic venoms or poisons; that is, the molecule attacks the nervous system. As discussed previously, the proteins embedded in the cell membrane of a neuron that transport sodium, potassium, and calcium are crucial to its overall function. Chemical agents such as snake venom and animal poisons, as well as insecticides, all either stimulate or block the activity of these proteins, interrupting cell-to-cell communication and ultimately causing death.

Pesticides, Rachael Carson, and *Silent Spring*

Given the similarities between the nervous systems of insects and those of vertebrate animals such as birds or mammals, it is not surprising that the wholesale spraying of DDT led to the immediate and significant killing of non-target animals. In an attempt to eradicate the gypsy moth from the northeastern United States, DDT, sometimes mixed with fuel oil, was sprayed from the air onto rural, suburban, and urban communities with abandon. Similarly, DDT was used in the control of Dutch elm disease, a fungus that is spread from tree to tree by the elm bark beetle. In the nineteenth and early twentieth centuries, elm trees became a highly desirable ornamental in urban environments and lined the median strips of major streets. Their density within such areas made them particularly susceptible to Dutch elm disease, which invaded North America from Europe. As elm trees died throughout the northeastern United States, extreme measures were taken to save the trees, including heavy application of DDT, which ultimately lead to dramatic mortalities of songbirds. The plight of these songbirds was the cornerstone of Rachel Carson's seminal *Silent Spring*, published in 1962.

It is not surprising that *Silent Spring* emphasizes the short-term exposure of wildlife to high concentrations of pesticides. First of all, the mortalities of songbirds, as stated above, were significant and undeniable. In 1962, there was no strong line of evidence to suggest that DDT (and the other recently synthesized organochlorine pesticides) would produce indirect impacts governed by the transport of the compound in the environment and its capacity for biomagnification through the food chain. The one major exception to this omission was the Clear Lake gnat study (discussed in chapter 7) that documented how the biomagnification of DDT subsequently led to the mass mortality of western grebes. To Carson's credit, in *Silent Spring* she not only highlights the results of the Clear Lake study but also evokes biomagnification as a mechanism leading to the nationwide decline she observed in the population of major predatory birds, such as bald eagles.

Soon after the publication of *Silent Spring*, the toxicological impacts of organochlorine pesticides, mitigated through biomagnification, became more evident. While the nationwide decline in bald eagles was being documented in the United States, a report out of Great Britain, published in 1963, confirmed that the breeding population of peregrine falcons was also decreasing, and that no falcons at all were breeding in southern England. It was noted that there was a correlation between DDT use and reproductive failure, and the British naturalist Derek Ratcliffe proposed that the two were causatively related. This led to an international conference regarding the population biology of peregrine falcons that documented a worldwide population crash of the species. While the conference did not reach consensus regarding the causative agent for the decline, pesticides, specifically DDT, were implicated.

The missing links between DDT, biomagnification, and declining populations of fish-eating birds and raptors began to become established in the late 1960s. In both the United States and Great Britain, studies regarding eggshell thickness produced the same results. On both continents, a sharp decrease in the weight of peregrine falcon eggshells was observed when eggs collected just before the adoption of DDT were compared with eggs collected after DDT had become widely used. The eggs of Long Island ospreys, as well as

Florida bald eagles, showed the same trend. Subsequent mechanistic studies suggested that, in raptors, it is the DDT metabolite DDE, not the parent compound itself, that is responsible for the eggshell thinning, through disruption of calcium transport within the eggshell gland.

While birds were showing pronounced impacts due to the biomagnification of organohaline pesticides through the aquatic food chain, was there a similarly consistent effect on a predatory mammal? Unlike birds, predatory mammals are rare in urban and suburban environments, and even when present they tend to be shy and retiring relative to charismatic and easily visible songbirds and raptors. Furthermore, unlike the fish-eating birds that are on the top of the food chain, coyotes, raccoons, fox, and skunks are omnivorous and can survive in the midst of humans and their associated pesticides without necessarily suffering from the maladies of biomagnification.

The noted mammalian exception, though, is the household cat, particularly cats that make their living in barnyards or other rural environments, eating rats, mice, and other vermin. Domestic cats groom with their tongues, and DDT that has accumulated on their fur can be ingested. For example, in 1965 an outbreak of hemorrhagic fever was apparently caused by a dramatic increase in the number of rats in a Bolivian community after the local cats had all died. Prior to the feline deaths, the houses within the community were sprayed with DDT, and it is likely that fur grooming caused the untimely death of the cats. Tissue from one of the dead cats was analyzed for DDT, and the concentration was indeed sufficient to be lethal.

In another example, this one from Borneo, the relationship between cats and DDT application takes an unusual turn that may or may not involve biomagnification. DDT had been sprayed on the longhouses of the natives for mosquito control. Domestic cats, reported to be at the top of a short terrestrial food pyramid, were supposedly dying from exposure to DDT through the consumption of geckos and roaches that lived in the thatched roofs of the longhouses. The death of the cats was reported to be associated with an explosion in the rodent population, which consumed stored crops.

The scarcity of cats in the villages ultimately led to a "cat-drop" in which felines were parachuted into the communities from cargo planes. Many aspects of the story have been documented, including the actual "cat-drop" itself. However, the biomagnification of DDT through a short terrestrial food chain, ultimately culminating in mortality within an apex predator, is supported only through personal narrative and hearsay, but not by any systematic scientific evaluation. As such, the biomagnification of DDT, through a terrestrial food chain, ultimately killing domestic cats is probably best understood as a folktale rather than an historical fact.

Humans have been plagued by, and have fought against, pests throughout our history. We have significantly raised the ante on pests at least twice during our history. The first time was with the advent of first-generation elemental and botanical pesticides, and the second, much more recent, was with the synthesis of organic pesticides. The production of second-generation pesticides did not occur in isolation but came about in concert with a number of other technological achievements. Organic synthesis generated a wonderland of chemical compounds, and the production of new chemical compounds continues unabated to this day. Many of the remaining chapters of this book focus on these newly formed chemicals and how they are transforming our future.

Chapter 13

The Origins of Regulation

If I have seen further it is by standing on the shoulders of Giants.
— Sir Isaac Newton

Toxicology textbooks are often structured to focus on one chemical class or another, or on one target tissue or another—a systematic approach that tends to overlook the social and political aspects of toxicology. Yet the story of modern toxicology, beginning immediately after World War II and continuing to the present, is a human-interest story as much as it is a story of polycyclic aromatic hydrocarbons and cancer, or psychoactive drugs and addiction. The next three chapters focus on public recognition of the potential harm caused by anthropogenic chemicals, and the efforts taken by a relatively few individuals to safeguard human health as well as the health of the environment. These key figures shaped contemporary views of chemical regulation, creating an important legacy for today's toxicologists.

The history of food, drug, and environmental safety regulation in the United States is one of hard-fought political battles that took years, and in some cases decades, to resolve. In many cases, singular events and charismatic individuals carried the day, galvanizing public opinion and legislative action. But before we delve into the tumultuous events following World War II, it is worthwhile to consider the state of food and drug safety in the United States during the nineteenth century.

In 1900, the population of the United States was 76 million, 60 percent of the population lived in rural environments, and one-fifth of them lived directly on farms. Many people's diets consisted of produce that they grew themselves, or that was produced locally. Meat was often from livestock grown locally, or from fish and game that were acquired locally through fishing or hunting. Ice was the primary means of refrigeration, and locally produced milk was unpasteurized. Microbiology was just emerging as a diagnostic field with respect to disease, and the control of diseases through sanitary food processing was in its infancy. Commercially processed foods were preserved through the use of a number of benign and nefarious chemicals, yet ingredient labeling on processed foods had yet to become a common practice.

The prospect of acquiring safe pharmaceuticals was no better than the prospect of acquiring safe food. Among "patent" medicines for common ailments, the more benign products often contained little more than color additives and sugar, whereas the more nefarious contained powerful narcotics such as opium, morphine, heroin, and cocaine. Bogus claims were made regarding the therapeutic capabilities of these medicines, and the public often mistook the psychoactive sensation of a narcotic for therapy. Medicine men competed with Wild West shows and circuses as a form of entertainment, with dubious medicines being hawked to the public as medical cure-alls. *Caveat emptor* was the watchword of the day, and customers had to look out for themselves as best as they could with respect to food and drug safety.

In the United States, the path from the relatively unregulated society of 1900 to where we are today was punctuated by key legislation that tightened national food and drug safety. Prior to the end of the Civil War in 1865, whatever regulations existed regarding food and drug safety did so under the purview of a specific state decree, and there was considerable variability, state by state, within the existing regulations. The post–Civil War period initiated a flourish of interstate commerce, which precipitated the need for a national agency to oversee food and drug safety. Furthermore, at the turn of the twentieth century, America accelerated its transition from a rural agrarian society to an urban/suburban one, further highlighting the need for centralized food and drug safety legislation.

The forebear of the Food and Drug Administration was the pure food movement, a grassroots effort that began in the 1870s. Support for the pure food movement originally came from the food industry itself as they inveighed against competitive new foods (oleomargarine competing against butter, cottonseed oil competing against lard) and also pushed for a unification of regulations that would supersede state-by-state rules. As one food packer said, "As it is now, we have to manufacture differently for every state." Despite the industry support, the movement was less an official coalition and more a movement created by different interest groups.

The federal precursor of the Food and Drug Administration was the Division of Chemistry within the Department of Agriculture. In 1883, the same year that the cholera bacterium was discovered by Robert Koch, Harvey Wiley became the Division's chief chemist. Wiley was a character, and he brought to the Division a number of publicity techniques that ran the gamut from unorthodox to downright bizarre. Wiley took the findings of his research to women's clubs as well as to civic and business organizations, and his efforts were rewarded with articles written in magazines such as *Collier's Weekly*, *Ladies Home Journal*, and *Good Housekeeping*. In addition, Wiley recruited twelve young men to volunteer and form a "poison squad." These intrepid volunteers agreed to eat only foods laced with known amounts of chemical preservatives such as borax; salicylic, sulfurous, and benzoatic acids; and the known chemical carcinogen, formaldehyde. The poison squad became an overnight media sensation, bringing even more media attention to Wiley and his cause.

While Wiley, working with President Theodore Roosevelt,

argued for the passage of legislation expanding the government's role in investigating food adulteration, stiff political resistance remained. The bill might not have made it out of the US House of Representatives if not for the timely publication of Upton Sinclair's novel *The Jungle* in 1906. *The Jungle* was intended to be a rallying cry against capitalism, as it promoted a more socialistic political system. However, the public's attention was most piqued by sections of the book that focused on the gross adulteration of foods with noxious chemicals such as formaldehyde, metallic salts, dyes, borax, and glycerine. Riding the political momentum created by Sinclair's book, Congress passed two major laws: the Federal Food and Drug Act, also known as "The Wiley Act," and the Meat Inspection Act. These acts established the Food and Drug Administration with a mission to "prevent the manufacture, sale, or transportation of adulterated or misbranded or poisonous or deleterious foods, drugs, medicines, and liquors, and for regulating traffic therein, and for other purposes."

From 1906 until the 1930s, the Food and Drug Administration focused on the confiscation of adulterated foods and drugs in order to keep them off the consumer market. The agency's role in regulating drugs or pharmaceuticals did not begin until the end of the 1930s, precipitated to a large degree by the Elixir Sulfanilamide tragedy. The history of sulfanilamide begins in 1932, when Gerhard Domagk discovered sulfamidochrysoïdine, which was marketed as Prontosil, the first commercially available antibiotic. It was later determined that Prontosil was metabolized in the body into the active ingredient para-aminobenzenesulfonamide, known also as sulfanilamide. The S. E. Massengill Company of Bristol, Tennessee, began to sell the compound, first in tablet form. To satisfy the demand for a liquid version, Massengill's head chemist sought a suitable solvent in which to dissolve the compound. He settled upon diethylene glycol, which sweetened the mixture. The solution was marketed as Elixer Sulfanilamide, despite the fact that it contained no alcohol; and no mention was made on the packaging that diethylene glycol was used in the formulation.

During the fall of 1937, 353 patients received and drank the elixir; of these, 105, or more than 30 percent, died. The alarming death toll led to an outcry from both chambers of the US Congress, as resolutions were passed requesting a full evaluation of the tragedy. The investigation led ultimately to the federal Food, Drug, and Cosmetic Act of 1938, which "legally mandated quality and identity standards for foods, prohibition of false therapeutic claims for drugs, coverage of cosmetics and medical devices, clarification of the FDA's right to conduct factory inspections, and control of product advertising, among other items." Functionally, the Food, Drug, and Cosmetic Act empowered the FDA to oversee the evaluation of new drugs, banning false and misleading labeling, and requiring formula disclosures on labels as well as directions for use and warnings regarding misuse.

World War II: The Century's Ultimate Paradigm Shifter

In the early 1900s, environmental regulation was on a relatively slow trajectory as it moved forward, motivated more or less, by one environmental disaster at a time. All of that changed with the advent of World War II, which brought about a series of major technological advancements with enormous implications for environmental regulation and human health. There were advances in medical and veterinary pharmaceuticals, the mass production of penicillin, and the agricultural and public health use of organochlorine pesticides. And of course, World War II marked the beginning of the atomic era, as the war was ended with the detonation of the first atomic weapons.

The research and development effort that produced the atom bombs, the Manhattan Project, was a clear testimony to the federal government's power of innovation. That level of innovation and the government's capacity to mobilize forces to achieve major goals was demonstrated in numerous other areas as well. One example was the boom in the production of a newly synthesized antibiotic, penicillin. Prior to 1945, the process of producing penicillin was a laborious, small-scale operation. In 1942, there was only enough penicillin produced to be administered to a total of twelve patients. By 1945, the process had become industrial, and production climbed to 4 million sterile packages of the drug per month.

At the height of the war, more than 1.9 billion people were in service on both sides of the war, either directly fighting, or fueling the war effort with other activities. Within the United States, the urgent necessities of World War II precipitated simultaneous innovation in a number of fields: operations research, electronic digital computing, radar, sonar, mechanized agriculture, communication systems, jet aircraft, rocketry, and, of course, medicine. While the use of antibiotics on the battlefield revolutionized medicine, pesticides revolutionized the control of infectious diseases. After the war, many of the changes imposed upon the agricultural, industrial, and social fabric of the United States were refocused toward the development of a postwar society. This included the development of the modern model for scientific inquiry, including the establishment in 1950 of the National Science Foundation.

The Postwar Toxicological World

World War II significantly influenced medical toxicology in two very important ways: the first was an intensified interest in cancer, while the second was a explosion of synthetic organic compounds onto the pharmaceutical marketplace. The 1945 detonation of two atomic bombs and the annihilation of two Japanese cities, Hiroshima and Nagasaki, made the horror of radiation poisoning devastatingly clear. After World War II, the specter of nuclear disaster loomed ever larger: the Cold War, atmospheric testing of atomic weapons from 1946 until 1963, and periodic accidents at nuclear reactors worldwide (beginning with the partial core meltdown at Chalk River, Canada, in 1952) all added to public fears of radiation poisoning. This anxiety about radiation and radiation-based carcinogenesis, in turn, helped to focus the nation's attention on cancer. Furthermore, the nationwide efforts that had successfully led to the mass production of antibiotics were now aimed at the study and prevention of cancer. However, as will be discussed in later chapters, despite the influx of funding, cancer proved to be a very tough problem to solve.

A second important consequence of World War II for toxicology was the burgeoning of the nouveau-chemical world. The flood of new pharmaceuticals, pesticides, and food additives onto the market greatly, and for the most part positively, impacted pharmaceutical production and medical practice. Nevertheless, it also led to dramatic and unanticipated side effects. In 1957, one of the newly synthesized chemicals, thalidomide, was registered in Germany for use as a sedative. The compound, marketed under the trade name of Contergan, was considered virtually nontoxic, as oral doses exceeding five grams per kilogram were not lethal to laboratory mice. Just two years later, European neurologists were diagnosing inflammation of peripheral nerve endings in patients who had been prescribed thalidomide, and that was just the tip of the iceberg. Two years later, in 1961, a sharp spike was observed in the occurrence of phocomelia, a development defect in which a major portion of the arms and legs of a developing fetus are missing so that the hands or feet attached to the body via small, irregularly shaped bones. The defect only occurred in children born to mothers who had been exposed to thalidomide during a specific period within pregnancy.

The association between thalidomide exposure and phocomelia was alarming enough that the manufacturer withdrew the compound from the European market in 1961. It has been argued that the real proof of the association between thalidomide and phocomelia was established only after the compound was removed from the market, as phocomelia virtually disappeared following its withdrawal. All told, during the four years that thalidomide was widely marketed, over 10,000 children were born with phocomelia.* By April 1962,

* An interesting sidebar to the thalidomide story in the United States is the role of an American medical officer, Frances Kelsey. Kelsey was in charge of the US Food and Drug Administration's thalidomide review and, acting according to established policy, she imposed a mandatory sixty-day waiting period prior to the drug's introduction onto the US market—during which time, as it happened, the alarm bells began to sound in Europe. As a result, the

the first animal study was published stating that thalidomide produced similar effects in the neonates of pregnant rabbits. Additional studies followed, and the fate of thalidomide's use during pregnancy was forever sealed.

The deformities caused by thalidomide were politically powerful enough to drive political change. In 1962, legislative change came in the form of the Kefauver Harris Amendment (known as the "Drug Efficacy Amendment") to the FDA's Food, Drug, and Cosmetic Act, an act that required, among other things, that drug manufacturers provide proof of the effectiveness and safety of their drugs before approval.

While the developmental abnormalities associated with thalidomide were immediate and obvious, more subtle damage from pharmacological exposure was also becoming evident. A classic example was exposure to the synthetic nonsteroidal estrogen, diethylstilbestrol (DES). DES was synthesized in 1938 and first prescribed to women in the 1940s and 1950s to prevent miscarriage. In contrast to the short shelf life of thalidomide on the pharmaceutical market, DES remained on the pharmaceutical market for twenty-four years, from 1947 and 1971. Furthermore, it has been estimated that between 5 and 10 million women in the United States and Europe were administered the drug.

Between 1966 and 1969, two decades after DES began to be prescribed, Dr. Arthur Herbst, a gynecologist at Massachusetts General Hospital (which was at the time called the Boston Lying-In Hospital), began seeing young women with a very rare cancer. His study followed eight women in their late teens and early twenties who had all been diagnosed with clear cell adenocarcinoma of the vagina. Interviews with mothers of the young women revealed an association between the disease and the administration of diethylstilbestrol during pregnancy. In 1971, Dr. Herbst and his colleagues published a

compound was never commercially available within the United States, and without this two-month safeguard, the history of thalidomide in the United States may very well have mirrored the European tragedy.

seminal paper in the *New England Journal of Medicine* describing the association between DES exposure and this rare cancer, speculating that early-term fetal exposure to DES was one of the risk factors for the disease.

In a follow-up study published four years later, Herbst and his colleagues compared 110 daughters from exposed mothers to 82 daughters from unexposed mothers. The women were examined for a number of benign alterations in the genital tract, including adenosis, or the presence of glandular tissue in the wall of the vagina. The study showed that the timing of exposure to DES was critically important. Of the women exposed to DES, almost three-quarters of them developed non-cancerous abnormalities when DES was administered to their mothers during the first two months of pregnancy. That incidence fell to approximately half of the DES daughters when exposure occurred during the third month of pregnancy, and fell to below one-tenth of the DES daughters when the exposure was initiated four months into the pregnancy.

The studies conducted by Dr. Herbst were responsible for a paradigm shift, fundamentally changing the way women and doctors thought about fetal exposure. DES, as it turned out, was altering pattern formation in the developing fetus in totally unanticipated ways, leading to an array of morphological abnormalities. Furthermore, while the initial studies were focused on cancer risk, the study conducted in 1975 was profound in that its focus turned from cancer to alterations in pattern development, even in cases where no cancers were observed in daughters of DES mothers.

These landmark developments are only part of the long, storied history of risk assessment and public health regulation. Often, safeguards have been instituted because of the actions of a large number of public health professionals—most working anonymously in the shadows of history. Nevertheless, singular paradigm-shifting events, such as those documented above and the individuals who have personified them, have left their mark on public health legislation over the last century.

Chapter 14

Low-Dose Chemical Carcinogenesis

Let me tell you something big: Give importance to little things!
— Mehmet Murat Ildan

At the same time that Frances Kelsey was acting as a human firewall, protecting the United States from the emerging European thalidomide tragedy, the United States was also struggling with the issue of how to regulate the use of chemical food additives. For the first half of the century, very few functional ingredients were added to food. After World War II, however, there was a revolution in food processing that rivaled the revolution that was occurring in the pharmaceutical and pesticide industries. The American way of life was becoming more urbanized, and there was public demand for foods that were of uniform quality, safety, and convenience. This demand, coupled with new scientific developments, led to a highly diverse supply of processed food, and led also to the use of a considerable number of food additives.

In the mid-1950s, in response to the burgeoning food additive market and the justifiable concerns that came along with it, New

York congressman Jim Delaney initiated a two-year investigation into the use of chemicals in food. The culmination of that investigation was the Delaney Committee Report. The report documented that food additives, in both the number of chemicals and the sheer volume used, were mushrooming. Further, the report noted that of the 704 chemicals used as food additives at the time, only slightly more than half (428) had been appropriately determined safe for consumption, and therefore the risk to the public was unknown and uncertain—which certainly did nothing to allay the concerns of the casual report reader.

The report garnered political support for legislative action, ultimately culminating in the 1958 Food Additive Amendment to the Food, Drug, and Cosmetic Act of 1938. The amendment required that chemicals used as food additives be tested for safety prior to their use, with the burden of proof being the responsibility of the industrial producer. While the amendment has been instrumental in the regulatory development of food safety regulation, it has also earned both fame and notoriety due to one specific passage, the so-called Delaney Clause: "*No additive shall be deemed to be safe if it is found to induce cancer when ingested by man or laboratory animals or if it is found, after tests which are appropriate for the evaluation of the safety of food additives, to induce cancer in man or animals.*" The Delaney clause appears in three separate parts of the amendment, in a section specifically on food additives, in another section related to animal drugs found in meat and poultry, and a third section on color additives. A fourth section, related to removing pesticide residues from foods, was also part of the amendment but was removed in 1996.

The Delaney Clause generated criticism almost immediately, as it brought a non-quantitative standard to a quantitative issue. At the time of the clause's adoption, only four substances were known to be carcinogenic: soot, radiation, tobacco smoke, and β-naphthylamine. Two developments occurred over the next twenty years that brought the problem with the Delaney Clause to the forefront. The first was a dramatic increase in food safety testing. To make a quantitative decision regarding the potential carcinogenicity of a chemical requires feeding high doses of a compound to laboratory animals (mice or rats), then extrapolating down to the low concentrations that humans experience when consuming processed foods. As more food additives came to market, more food testing was being accomplished. The second development was a dramatic increase in the analytical capability to detect specific compounds in food. Between 1958 and 1980, the ability to quantitatively measure specific compounds in food increased by between 100- and 10,000-fold, depending upon the specifics of the chemical.

This one-two combination revealed some difficult realities. Carcinogenic compounds, in minute concentrations, were turning up in common foods that had been considered safe for decades. Many of these compounds had been tested for food safety and regarded as safe, but were now violating the Delaney Clause. To avoid chaos, the food additive amendment included the concept of GRAS: Generally Regarded As Safe. If a chemical had been tested previously and was regarded as safe, then it would not be banned—even if it contained trace levels of carcinogenic compounds.

The term "generally regarded as safe" may sound arbitrary, but it was not. The beauty of the rodent testing was that it produced very precise dose–response relationships that were used to establish probabilities of carcinogenicity. Through the wonders of mathematical extrapolation, tumors in rats that had been fed high doses of a food additive could be used to extrapolate the probabilities of low doses generating cancer in humans, and the Delaney Clause included an escape clause, known as *de minimus*. For a compound to be generally regarded as safe does not need to imply that it is not carcinogenic at all, but rather that the risk to humans of developing cancer within one person's lifetime is less than one in a million. Consequently, the application of GRAS and *de minimus* allowed for an end run around the Delaney Clause, putting a quantitative limit on the declaration of carcinogenicity despite the fact that this was effectively violating the letter of the law of the Delaney Clause. The law had been a stringent attempt to limit carcinogenic compounds from entering the food supply, but at the end of the day, regulators turned back to the old standby, the dose–response curve. However, even the old standby was having its problems.

A classic example of regulation winding up in a quagmire was the case of the artificial sweeteners cyclamate and saccharin. Saccharin fell under the regulatory microscope soon after its synthesis in 1878. Harvey Washington Wiley, of the original Wiley Act of 1906, wanted it banned, but the corpulent President Theodore Roosevelt had been using the chemical for weight control for years. The two clashed. Wiley is reported to have told the president that eating saccharin was ". . . eating a coal tar product totally devoid of food value and extremely injurious to health." Roosevelt replied tartly, "Anybody who says saccharin is injurious to health is an idiot." Soon thereafter Wiley's political career was finished, while saccharin remained on the market. Saccharin ultimately received GRAS status in 1958.

In 1937, fifty-nine years after the synthesis of saccharin, cyclamate was synthesized and, in 1950, it went into production. Despite being on the market for only eight years, cyclamate also received GRAS when the Delaney Clause became law. Saccharin and cyclamate were often used together, in a 10:1 cyclamate-to-saccharin blend, as the bitter taste of saccharin was counteracted by the addition of cyclamate. In the late 1960s, a series of studies were conducted on cyclamate, questioning its safety. The *coup de grâce* came from a study linking the cyclamate/saccharin mixture to bladder cancer in rats. In 1970, cyclamate was banned from use for violating the Delaney Clause. Subsequent research failed to substantiate the claim that cyclamate caused bladder cancer, however, and in 1984 the FDA asked the National Academy of Sciences to review the conflicting literature. As a result of that review, cyclamate was cleared of being carcinogenic but maintained the status of being co-carcinogenic or in other words, a substance that promotes the action of other carcinogens. Despite the recommendation of the National Academy of Sciences, the FDA refused to remove the ban, and cyclamate remains banned from the marketplace to this day.

The fate of saccharin was completely different. After cyclamate had been removed from the market in 1970, saccharin was the only artificial sweetener left standing. The FDA asked the National Academy of Sciences to conduct another review of the safety of saccharin, and while the resulting report determined saccharin to be safe, the Academy recommended further study. The result from those additional studies: an association between saccharin and bladder cancer. The indictment was not completely clear, however. While seven different lab groups conducted studies with saccharin, only two identified bladder tumors. Furthermore, the dose at which bladder cancer occurred was very high, the human equivalent of some 870 cans of soda per day. This represented an impossibly high level of exposure, but recall how regulatory agencies use dose–response curves. Very high levels of a chemical are administered, followed by statistical extrapolation back to "normal" levels, with a quantitative probability put on the chances of consumption, at normal or reasonable levels, leading to carcinogenicity. High levels of exposure, while seemingly ridiculous when put into the context of daily intake, are actually part and parcel to the regulatory use of the dose–response curve.

The Delaney Act could have been used to ban saccharin, just as it had been used to ban cyclamate, and the then commissioner of the FDA, Charles C. Edwards, said as much. However, with the removal of cyclamate from the market years before, saccharin was the only artificial sweetener on the market. The American public had developed a fondness for artificial sweeteners, and the political blowback from an all-out ban of saccharin would have generated too much political pressure for it to occur. For a time, a compromise position was reached: foods containing saccharin were labeled "Determined to cause cancer in laboratory animals." However, as further evidence was brought to bear, the bladder cancer in rodents was not considered to have an analogous etiology in people, and in 2000 the warning labels on saccharin packets were removed.

Cancer Initiation, Promotion, and the Low-Dose Effect

As stated in chapter 10, one method by which chemical carcinogenesis can be initiated is via the formation and dysrepair of DNA adducts, leading to genetic mutation and ultimately to tumor formation. Once a cell has been mutated by an initiator, it becomes susceptible to the action of chemical promoters that stimulate tumor growth. In some cases, the intiating and promoting chemicals are caused by the same compound; however, in many other cases the two compounds are different.

Understanding the activity of chemical cancer promoters necessitates research on living cells that have been harvested from cancer patients. Early studies featuring cancer cell lines were problematic, as the harvested cells did not last more than a few months in culture. The cell lines would divide and flourish for a few months, then senesce and die out. Without an immortal cell model, considerable effort was put into harvesting new cells from different patients, and the cell lines were all so genetically different that a comprehensive understanding of any given cancer was difficult to stitch together.

Research into the role of cancer cell promoters was restricted by these difficulties with cell lines until immortal cell lines were established. An *immortal cell line* is one that has experienced a mutation such that the genes associated with senescence, or biological aging, have been silenced. Consequently these cell lines, when properly cared for, never die. The technological ability to establish immortal cell lines was perfected in the early 1970s. One important cell line was developed by the Michigan Cancer Foundation from a woman diagnosed with what would ultimately become terminal breast cancer. That woman was Sister Catherine Francis Mallon, a nun living in the Detroit area, and the cell line, which became known as MCF-7, is highly estrogen receptor–positive, meaning that the common estrogen, 17β-estradiol, promotes cell proliferation.

The development of the MCF-7 cell line, as well as others, ushered in a bustling period of research to unlock the mechanisms of estrogen-based carcinogenicity. In 1975, scientists demonstrated that the antiestrogen tamoxifen inhibited the growth of MCF-7 cells. This finding had profound implications, as tamoxifen, which competes with estradiol to bind to the estrogen receptor, subsequently went on to become a standard component of therapy for virtually all cases of estrogen receptor–positive cancers. What was not known was how estrogens lead to cancer and whether or not the steroid itself was carcinogenic.

During the mid-1970s and into the 1980s, the central question was how estrogen led to tumor formation, and the debate was driven by conflicting results. In some laboratories, estrogen led to a proliferation of MCF-7 cells, while in other laboratories it did not. The resolution of the issue was startling: in the laboratories in which estrogen did not cause cell growth, the cell line was already stimulated by an estrogenic substance prior to exposure. The commercial media on which the cells were grown contained a chemical, phenol red, that was capable of binding to the estrogen receptor, thereby stimulating cell proliferation. Without phenol red in the media, estradiol's role in cell proliferation was obvious. With phenol red in the media, even at low concentrations, estrogen could not stimulate cell growth because that mission had already been accomplished by the interloping endocrine-disrupting compound, phenol red.

The toxicity of phenol red does not behave according to the constructs of a simple dose–response curve. At high doses the compound is toxic, killing estrogen-responsive and non-estrogen-responsive cells alike. At levels below this overt toxicity, the compound does not kill but actually acts as a growth promoter, stimulating estrogen-responsive cells to divide. While this might seem benign, in the case of MCF-7 cells it decidedly is not, for the cells that are being stimulated to grow are carcinogenic; that is, they are tumor producers, and phenol red is acting as a cancer promoter. The response may still be toxic (to the cancer patient if not the cells themselves); however, the impact on the cell line has changed from outright cell death to the stimulation of cell growth. At much lower doses, the impact of phenol red as an promoter of tumor cell growth attenuates as the concentration of the compound is insufficient to adequately stimulate the growth response.

The confounding impact of phenol red on the growth of MCF-7 cells was surprising, but subsequent results have illustrated that the impact of rogue chemicals on cell growth is not an isolated event.

For example, one of the developed assays for estrogenicity involves growing MCF-7 cells in the presence of human serum that has been stripped so as to be steroid-free. Growth of the cells in this environment is inhibited; therefore estrogens are added to the culture to rescue the cells from inhibition, allowing them to proliferate. When Ana Soto's laboratory at Tufts University applied this technique to cells grown in a specific brand of culture tube, they could not reproduce the results. A thorough analysis of the chemical milieu in which the cells were growing revealed the presence of a rogue estrogenic compound, p-Nonyl-phenol, that was leaching from the plastic of the culture tube into the solution in which the cells were growing. Because the p-Nonyl-phenol was stimulating the cells to grow, the cultures in which growth was supposed to be suppressed (the negative controls) were experiencing unexpected cell growth.

The two examples above illustrate an important characteristic of steroid receptors—an alarming lack of fidelity. Studies reveal that a large number of chemicals are able to bind to these receptors. Even ions of the element cadmium bind to the steroid receptor, causing estrogenic effects, despite the atom's diminutive size. Part of the reason for the receptor's promiscuity can be traced back to its evolutionary family origin. The steroid receptors belong to a larger group of receptors, the nuclear receptor superfamily. Over 300 members of this superfamily have been identified, and many are orphan receptors, having no known natural endogenous ligand. These receptors lead to the production of Cytochrome P450 enzymes, those enzymes involved in the biotransformation of exogenous chemicals from lipid-soluble to water-soluble, which allows for efficient excretion (chapter 5). The steroid receptors are members of a family that can respond to a suite of chemicals, thereby reducing their toxicity. Given that the estrogen receptor is a member of this family, its lack of fidelity, rather than being surprising, is perhaps to be expected.

A second attribute of steroid hormone receptors also contributes to its ability to respond to very low hormonal signals. Hormones such as 17β-estradiol circulate in the blood and then bind to cellular receptors at very low concentrations. For example, while blood glucose levels are normally found at a level of about one part per thousand, or a gram per liter, circulating levels of steroids are much lower than that, about a million times lower, circulating in the blood at parts per billion levels. The endocrine system has evolved to amplify this very quiet signal by having the molecule bind to a specific receptor that has a very high affinity for it.

Importantly, the relationship among the concentration of steroid in the blood, the occupancy of receptors, and biological effect is not linear. At very low doses, an increase in steroid concentration can be matched by a consistent increase in the percentage of the total number of receptors that are occupied. At higher doses, the increase in steroid concentration does not lead to a consistent increase in the percentage of steroid-bound receptors; rather, the increase in the rate of steroid binding to receptors declines.* But there is more to the story. For many biological effects in which a ligand such as a steroid has to bind to a receptor, the relationship between binding and biological effect is straightforward, such that the highest levels of bound receptors (approaching 100 percent bound) elicit the greatest biological impact. But not so with steroids—a near maximal biological effect can be elicited well below the point at which all of the receptors are bound to the steroid. This is known as the *spare receptor hypothesis*, for even if only a small number of receptors pick up the steroid signal, the biological response is maximized.

This capacity to amplify these minute chemical signals comes with a cost, for the steroids associated with reproduction come with consequences. Even as far back as the 1930s, it was known that rodents exposed to elevated levels of estrogen, the primary feminizing sex steroid, develop mammary cancer. However, until recently, the mechanism by which steroids were linked to cancer was unclear. Elucidation of estrogen as a cancer promoter would have to wait for the development of the immortal cell lines, as explained previously.

* If the relationship between the concentration of steroid and the occupancy percentage of a steroid receptor is plotted, the result is the well-described sigmoidal dose–response relationship discussed at length in chapter 1.

Over the last fifty years, we have developed a better understanding of the environmental impact that chemicals have on humans and the environment. While we have long recognized overt and acute toxicity, the field of toxicology has broadened to include subtle effects brought about by lower-level but chronic exposures. We are no longer focused solely on fatality, but on less obvious consequences such as carcinogenicity and the embryological effects that occur in developing animals. In highly visible cases, such as that with thalidomide, the impacts are immediately recognizable, and the lag time between initiation of the impact and overt manifestation is short. In more subtle impacts, as will be highlighted in some of the subsequent chapters, chemicals can target internal organs and the impacts may not be readily recognizable at all.

Chapter 15

POPs and Silent Spring

We spray our elms and the following springs are silent of robin song, not because we sprayed the robins directly but because the poison traveled, step by step, through the now familiar elm leaf–earthworm–robin cycle.
— Rachael Carson, *Silent Spring*

While food and drug safety regulations in the United States were moving forward one regulatory act at a time, the environmental movement had yet to become mobilized. The US frontier closed in 1890 and up until that time, and for some considerable time past it, the vastness of the American landscape made the premise of pan-continental contamination seem preposterous. The voices speaking about environmental issues at the turn of the twentieth century were wilderness enthusiasts such as President Theodore Roosevelt, his chief advisor, Gifford Pinchot, and John Muir, founder of the Sierra Club. To put the era into perspective, Yellowstone National Park—the nation's first—was created in 1872, just twenty-eight years before the turn of the century. The prospect that chemicals could alter the environment to such a degree that they could effect the entire North

American continent was still decades away. Even after the official closing of the Western frontier, the vast geography of the American West appeared to be infinite.

As was true for food and drug safety, World War II proved to be a pivotal moment for the environmental movement. Industrial mobilization during the war not only affected the production of drugs such as penicillin, but also the production of pesticides such as DDT. In 1943, the production of the insecticide was negligible. Given its importance as a control agent for diseases such as malaria, where insects are the primary vectors, DDT production was rapidly ramped up during the final years of the war. By 1945, production increased to over 2 million pounds of DDT synthesized per *month*. When the war ended, increases in the production of organochloride pesticides continued virtually unabated.

By the early 1960s, marine biologist Rachael Carson began to write *Silent Spring*, which is now considered a foundational text of the environmental movement. Carson found herself writing in a very unsettled social and political climate. Postwar Americans were increasingly anxious about radiation, cancer, and the specter of nuclear war. The atomic bombs dropped on Japan, and the inevitable instances of cancer caused by those bombings, as well as the threat of all-out war with the Soviet Union, weighed heavily on the American psyche. Rachel Carson understood this, and in her writing she linked the fears associated with radiation to the issue of pesticide overuse. Tellingly, in a book that focuses on chemical contamination, the word *radiation* appears no fewer than 50 times, while the word *cancer* appears in the book over 110 times. *Silent Spring* was able to speak to the societal angst associated with radiation poisoning and direct it toward the burgeoning, widespread use of pesticides.

The pesticides that Carson specifically discussed were three organochlorides: DDT, aldrin, and dieldrin. These pesticides are part of a larger group of toxic compounds known as *persistent organic pollutants,* or POPs. POPs are lipophilic and highly resistant to biotransformation; they will seek out a lipid source in the environment, most notably organic sediments or organisms themselves. Once inside an organism, POPs tend to accumulate in fat cells, where they remain essentially unchanged for long periods of time. When these organisms enter the food chain and are eaten by other animals, the POPs biomagnify. This biomagnification leads to very high concentrations of POPs in large raptors, such as eagles and ospreys, birds that occupy the top of many aquatic food chains, and it was the decline in these charismatic predatory birds that drew the attention of Rachael Carson and others. Ultimately, the wholesale disappearance of raptors became a rallying issue for the burgeoning 1960s environmental movement. In a manner similar to the early epidemiological studies conducted by Percivall Pott and others, ecotoxicology was born by chronicling a highly visible occurrence, the decimation of top predatory birds.

At the time of Rachael Carson's writing of *Silent Spring*, the concept of biomagnification was unexpected for at least two major reasons. The first was that, at the time, toxicology focused on acute responses of animals to high concentrations of toxic compounds that led to an immediate adverse toxicity, possibly including death. Using wild animals to evaluate long-term adverse ecological impacts was in its infancy during Rachel Carson's time. Even the use of laboratory animal models to confirm chemical carcinogenicity only preceded *Silent Spring* by about forty-five years. In those studies, rats, mice, and rabbits were directly exposed to high doses of carcinogens for long periods, but even with such overt exposures it had proven difficult to establish links between chemicals and cancer. For wild animals to become harbingers of environmental dysfunction, they would literally have to begin to fall out of the sky. Which, in the example of predatory birds and DDT, is exactly what they did.

The second reason why adverse impacts from biomagnification were unexpected was the adage that "the solution to pollution is dilution." In classic exposures to toxic compounds, a direct exposure to the chemical source is necessary. Should the animal survive the exposure, sequestration (in an inert, benign form), or metabolism, or excretion would have been expected to reduce the chemical concentration to background levels. For the chemical to actually *increase* as it was "diluted" through the process of food chain consumption just didn't seem possible.

Silent Spring took the nation by storm. By 1964, the year of Carson's untimely death, the book was a best seller, with more than a million copies sold that year. When pressed, then President John Kennedy professed an interest in the book, ultimately appointing a Presidential Science Advisory Committee to look into its findings. The committee confirmed Carson's allegations that the indiscriminate use of pesticides was causing profound environment effects through biomagnification of the compounds through the food chain. Her work proved to be the death knell for the three pesticides that she made infamous, as DDT was banned for use in the United States in 1972, and both aldrin and dieldrin were banned in 1974. Furthermore, *Silent Spring* also provided a catalyst for significant US legislation: the Clean Air Act of 1963, the establishment of the US Environmental Protection Agency in 1970, and the Clean Water Act of 1972.

The story of POPs might very well have ended with *Silent Spring* if not for one more curious property associated with this group of compounds: they travel. This realization began to spread around the globe almost immediately after *Silent Spring* was published, although it was brought to the spotlight not by a POP, but rather by coal, a familiar toxicological nemesis. In the 1960s, coal-burning power plants generated considerable amounts of sulfur emissions throughout continental Europe. During the 1970s, studies were conducted demonstrating the vast distances that these air pollutants could travel: many thousands of miles. The traveling distances strongly suggested that international cooperation was necessary to adequately tackle problems associated with air pollution, and the need for such cooperation precipitated the 1979 Geneva Convention on Long Range Transboundary Air Pollution.

In fact, it wasn't just sulfur dioxide that was traveling, but also some metals and the POPs. In 1998, members of the United Nations Economic Commission for Europe signed the legally binding Protocol on Persistent Organic Pollutants as an addition to the 1979

Convention. The protocol focused on eliminating the discharge and emission of sixteen POPs to the environment, including eleven pesticides, two industrial chemicals, and three industrial by-products. It banned the production and use of some POPs (aldrin, dieldrin), created a schedule for the elimination of others (hexaclorobenzene, PCBs), and restricted the use of still others (DDT). The Convention was expanded internationally in 2000 when diplomats from 122 countries completed a treaty controlling POPs. The treaty was signed into international law in 2001 at the Stockholm Convention on POPs. Implementation of the law focused on twelve compounds, including DDT, aldrin, and dieldrin, which were collectively dubbed the "Dirty Dozen." Since that time, additional compounds have made the list, with the criteria focusing on persistence, the capacity to biomagnify through the food chain, the capacity for long-distance atmospheric transport, and the capacity to adversely impact human health.

The term *persistent organic pollutant* does not appear in *Silent Spring*, nor was Carson able to predict that these compounds could travel great distance. Her indictment of pesticides is based upon acute, high-dose exposures leading to rapid death, the type of response that is most evident to the casual observer. Carson did not anticipate that these compounds could travel and disperse globally in more or less their parent form. In fact, she states in the book,

> When some of the Eskimos themselves were checked by analysis of fat samples, small residues of DDT were found (0 to 1.9 parts per million). The reason for this was clear. The fat samples were taken from people who had left their native villages to enter the United States Public Health Service Hospital in Anchorage for surgery. There the ways of civilization prevailed, and the meals in this hospital were found to contain as much DDT as those in the most populous city. For their brief stay in civilization the Eskimos were rewarded with a taint of poison.*

Carson did not even consider that the Arctic natives might not have had to travel to Anchorage for a dose of DDT; the DDT may very well have traveled to them.

---

* Rachel Carson, *Silent Spring* (New York: Houghton Mifflin Company, 1961), 179–80.

Just how does this traveling occur and why would the compounds wind up in the most remote areas of the world, as opposed to the most populous or the most polluted? The answer is known as global distillation, or the "grasshopper effect." The POPs are semivolatile, meaning that they vaporize and enter the atmosphere when the weather is warm. When temperatures cool, the compounds condense, drop out of the atmosphere, and fall back to earth. Migration of the compounds from one region to another is a slow process, involving multiple vapor/condensation cycles, and consequently the grasshopper effect only occurs with compounds that are particularly resistant to degradation, such as POPs. If the compound was to enter a region where it is always cold, such as in the Arctic or Antarctic, it would remain there, effectively locked in its condensed form, unable to return to the atmosphere and continue its wanderings around the globe. Contamination of the Arctic by POPs represents a perfect storm, as the chemical properties of POPs make it likely that they will aggregate in the far North, accumulating in top predators: marine fish, birds, and mammals. The grasshopper effect allows them to travel around the globe, while their resistance to degradation prevents them from disappearing in the process. Once a compound reaches the polar regions, its lipophilicity dictates that it enter the biota rather than remain in the water. Lipophilic contaminants bioconcentrate in organisms, and will also biomagnify as they move up the food chain into the top predatory animals.

Another factor facilitating the accumulation of POPs in the Arctic is the Arctic ecosystem itself. The ecosystem is characterized by low levels of plant productivity, low levels of species diversity, and relatively simple food chains. The Arctic food chain begins with plankton, the single-celled plants that photosynthesize and fix carbon into more complicated organic compounds. These single-celled plants are consumed by zooplankton, a diverse array of tiny animals that includes copepods and amphipods. The small shrimp-like creatures, in turn, are consumed by fish, including the Arctic cod—one of the most abundant marine fishes in the Arctic and an important prey item for marine mammals. This short-lived fish resides under the ice, where it consumes amphipods and copepods, concentrating POPs within its system as it eats. Arctic ringed seals consume fish, including the Arctic cod, and the seals in turn are preyed upon by the top predator—the polar bear. Dramatic biomagnification of PCBs up the food chain have been documented as concentrations of PCB congeners increase in fatty tissue of the animals within the cod-seal-bear food chain—from 0.0037 to 0.68 to 4.50 mg/kg, respectively.

In Arctic marine mammals, POPs tend to be more concentrated in males than females. For Arctic ringed seals, the males have been shown to contain twice the levels of DDT and PCBs as do the females, whereas the concentration of PCB cogeners in the fat of polar bears was 50 percent greater in males than it is in females. Humans indigenous to the region experience the same trend. When the blood of Inuit living on Broughton Island, Northwest Territory, Canada was tested for PCBs and dioxin-like compounds, it was found that the concentrations actually increased with age and were greater in male than female subjects.

The escape clause for females is lactation, the production of fatty milk; however, while lactation allows the concentration of POPs in the females to decrease, it simultaneously mainlines POPs to the developing juveniles. Breast milk from Inuit women, consuming "country food" (traditional foods collected from the ecosystem), contains almost twice the POP contamination as that of women in modern settlements eating market food. Breast milk of Inuit women was consistent with that of beluga whales, and was seven times greater than that within the fatty tissue of Arctic char. It is not a stretch to say that the top of the Arctic food chain is inhabited by mammalian infants, the milk drinkers, be they polar bear, toothed whales, or humans.

The main focus of *Silent Spring* is clearly the overuse of organochlorine pesticides, those highly mobile POPs that travel the globe in search of a cold, fatty resting place. But the story of POPs is only part of the genius of the book, for *Silent Spring* is also a bellwether, ushering in the modern era of ecotoxicology. While Carson misinterpreted the meaning of POPs in Inuit peoples, *Silent Spring* opened the public's eyes (including those of young and upcoming researchers) to the fact that trends within the natural world can be viewed as harbingers of impending environmental dysfunction.

The publication of *Silent Spring* corresponded with a profound paradigm shift in ecotoxicology. Rather than being a benign influence, humans were, perhaps for the first time in history, clearly shown to have the capacity to alter entire ecosystems through the excessive release of industrial chemicals into the environment. Furthermore, by conducting a meta-analysis (an analysis of compilation of the results from many widely disparate studies), Carson was able to articulate these disparate bits of information, and formulate them into her transformative, though chilling, conclusion. Within the text, Carson touches upon such forward-looking topics as transgenerational toxicology, reproductive toxicology, the use of environmental sentinel organisms, the prospect of chemical mixtures in the environment causing unexpected health impacts, elevated exposures within indigenous peoples from polar regions, differential sensitivity to pesticides, and endocrine disruption through the lens of hormonally based carcinogenesis. While the end of World War II ushered in the modern chemical age, *Silent Spring* ushered in the modern age of environmental awareness of the consequences of modern chemicals. From an environmental perspective, current approaches toward environmental toxicology can all be traced back to this one seminal book.

Chapter 16

Toxic Toiletries

We have found the enemy and he is us.
— Pogo, from the comic strip *Pogo* by Walt Kelly

At the turn of the century, between 1999 and 2000, Dana Kolpin and a small group of his fellow United States Geological Survey scientists were busy. Over the course of two years they conducted a nationwide reconnaissance to measure organic wastewater contaminants in 139 streams in thirty different states. The water they collected was analyzed for a variety of pharmaceuticals, including veterinary and human antibiotics, prescription and nonprescription drugs, as well as steroids and hormones. They also tested for a few personal care products, including DEET (N,N-diethyl-m-toluamide), the primary chemical used in many common insect repellants, and triclosan, an antibacterial compound found in soaps and detergents. While the sampling sites were biased toward streams adjacent to potential sources of contamination, and most chemicals were found in only very low concentration (less than one part per billion), the results were still striking. Fully 80 percent of the 139 sampled streams contained detectable levels of a number of compounds found, on average, in a mixture of seven different chemical compounds per site.

Many of the chemicals identified by Kolpin's group fell into the category of *pharmaceuticals and personal care products* (PPCPs), a classification that, until recently, was not considered to be toxicologically relevant. The results of this research turned many heads; since there was no water-quality criteria for many of these chemicals, the risk associated with their presence in surface waters was unknown. Furthermore, while concentrations were low, the nearly ubiquitous presence of the chemicals in the environment provoked scientists to reconsider their potential toxicity and to take a closer look at their overall life cycle, from production to personal use to ultimate disposal.

The US Food and Drug Administration divides PPCPs into two categories: cosmetics and drugs. Cosmetics are intended to cleanse or beautify, while drugs are intended to diagnose, cure, mitigate, treat, or prevent disease. Sunscreens and acne creams affect the structure and function of the body; therefore they are also classified as drugs. Some products, however, such as moisturizing sunscreens or anti-dandruff shampoos, blur the distinction between the two classifications and are classified as both.

The current use of pharmaceuticals is staggering. More than 4 billion medical prescriptions were dispensed in the United States in 2013, with more than 45 percent of all residents having been prescribed at least one pharmaceutical compound every month. The use of nonprescription pharmaceuticals is equally impressive; for example, the annual consumption of aspirin in the United States exceeds 10,000 tons.

The use of ingredients within personal care products is a bit harder to track, as they are formulated mixtures of compounds, rather than single chemicals, and the exact formulation of a product can be proprietary. As such, the specific chemical composition within a product, and therefore the exposure dose of the chemical to the individual using it, can be difficult to estimate. However, when considered in terms of the number of personal care products used daily, the number is striking. In a recently conducted survey conducted by the Environmental Working Group, women reported using an average of nine different personal care products every day, while 1 percent of all men and 25 percent of all women reported a daily use of fifteen or more different personal care products. Personal care products include such everyday products as lip balm, colognes, deodorants, perfumes and lotions, makeup and eye liner, shaving cream and skin cream, cleansing pads, cotton swabs and pads, toilet paper, facial tissue, and wet wipes, among other products. Within these products, there are over 10,500 different unique chemical ingredients. Furthermore, many of these compounds are not found in only one product, but rather they are distributed among a wide variety of products. For example, the antibiotic triclosan is found in a number of different antibacterial soaps, toothpaste, lip gloss, first-aid creams, deodorants, and even certain types of kitchenware, mattresses, and children's toys.

Production and Usage of PPCPs

The large demand for PPCPs has spawned a rapid development in the chemical industry that supplies them. For pharmaceuticals, the production pathway invariably leads overseas, as over 80 percent of the active pharmaceutical chemicals being consumed in the United States each year, as well as 40 percent of the finished product, are being manufactured internationally. The primary countries that are now in the business of satisfying the US market's pharmaceutical needs are India and China, and their enthusiasm for the task is saddling these countries with significant water pollution issues.

Perhaps the most striking example of excessive environmental release of pharmaceuticals is occurring in the town of Patancheru, within the Hyderabad region of India. This region has become a focal point for India's pharmaceutical industry, as more than ninety pharmaceutical plants are located nearby. As is true with many chemical processes, an increase in production can lead to a concomitant increase in the loss of product in the waste stream. Consequently, the effluent from the pharmaceutical plants surrounding Patancheru has to be treated prior to its discharge into surface waters. Unfortunately, the water-treatment facility charged with removing these chemicals from the effluent water, the treatment plant run by Patancheru Environ Tech Ltd., is not up to the task, and consequently the wastewater leaving the treatment facility contains pharmaceuticals in very high concentrations. Eleven pharmaceutical compounds, including six antibiotics, a blood pressure regulator, four chemical receptor blockers, and a serotonin reuptake inhibitor have been found to occur in the wastewater at concentrations exceeding 100 micrograms per liter. These concentrations are significant, as they all exceed the toxicity values for aquatic biota that may be exposed to them. The compound with the highest-recorded levels in the wastewater is the antibiotic ciprofloxacin, with samples ranging up to 31 milligrams per liter, a concentration that exceeds the therapeutic dose of the chemical when medically administered. The situation in China is similar, as levels of the antibiotic oxytetracycline were found in final treated wastewater effluent, as well as the receiving waters, in parts per million (milligrams per liter) concentrations.

The pathway from production of chemical to final product in a home medicine cabinet can vary substantially from one PPCP to the next. While some products in their sellable form may reach the United States from overseas producers, the ingredients of others are produced abroad while the final product is formulated within the United States, then packaged and sold to US consumers. Regardless of the intermediate steps between chemical producer and consumer, virtually every US citizen is a consumer of PPCPs.

The safety of personal care products (PCPs), like drugs, falls under the purview of the US FDA, although the laws are more lax for PCPs than they are for drugs. The law prohibits the marketing of adulterated or misbranded cosmetics, but it does not give the FDA the authority to require pre-market approval of PCP ingredients. The one exception to this law focuses on color additives, which do require pre-market approval. It is not the FDA, but rather the manufacturers themselves that are legally responsible for the safety of their products.

One poignant example of the inadvertent toxicity of a class of PCPs focuses on certain hair products, including hair straighteners. These products contain extracts from animal placentas, which in turn contain progesterone, estrogens, and growth factors, all of which are bioactive. Furthermore, such products are generally applied to the hair, hair follicles, and scalp for prolonged periods of time. The growth factors have been shown to increase hair follicle growth and decrease hair shedding, most likely by increasing blood vessel formation and subsequent blood flow to the hair follicles. These products, however, are also associated with a number of adverse health outcomes, including the initiation of menarche (the first menstrual cycle) in girls at earlier ages, and a higher risk of uterine leiomyomata (benign smooth muscle tumors that only rarely become cancer) in adult women. Perhaps the most disturbing outcome occurs when these products are used on young girls. In children as young as fourteen months old, hair products have been found to cause premature development of secondary sexual characteristics (breast development and pubic hair growth). Fortunately, once application is stopped, children revert to their age-appropriate stage of development.

While the hair products mentioned above may be extreme examples, many PPCPs are designed to be biologically active compounds, meaning that they are specifically designed to elicit cellular change. Steroids, antibiotics, and prescription and nonprescription drugs are all biologically active compounds, designed to interact with specific metabolic pathways and processes that occur within the target animal or person. Personal care products such as antibacterial soaps or toothpastes may also be deliberately designed to be biologically active or may inadvertently contain biologically active ingredients, even when they are not designed for that purpose. Regardless of the intent, PPCPs induce cellular or molecular changes and this is where the problem lies; not all of the changes are necessarily beneficial, nor are they limited to the user of the product.

PPCPs in the Environment: Unintended Consequences

After their use, drugs are excreted from the body, most often in the urine, while personal care products, applied locally, are washed off. In both cases, they can enter the wastewater stream either in their original form or with molecular side groups attached to make them more water-soluble. Importantly, these water-soluble metabolites do not necessarily lose their biologic potency, but rather can remain biologically active as metabolites or can be reconverted (thorough the action of bacteria) back into the original compounds that are then available in the environment at their full strength. This is where Kolpin and his USGS colleagues reenter the story of PPCPs. The fact that these compounds are found so ubiquitously across the country, coupled with the fact that they are found in complex mixtures of seven or more compounds at each site, highlights PPCPs as potentially important new environmental contaminants.

Given concerns about the environmental safety of these compounds, is there evidence that they interfere with local biota? The answer is clearly yes, as is evident from an example that comes from Asia, and concerns the relationship among the Oriental white-backed vulture (*Gyps bengalensis*), domestic cattle, and the anti-inflammatory drug diclofenac.

Vultures are predatory scavengers, feeding on the carcasses of dead animals. At one time, the white-backed vulture was the most abundant predatory bird in the world, and its success was due, in large part, to its close association with humans and their livestock. In India, cattle have traditionally been used to produce milk and as beasts of burden, but are not eaten. When one of the country's 500 million head of cattle dies, the preferred method of disposal is consumption by vultures, even in cities.* In the 1990s, the number of vultures began to decline noticeably, culminating in the loss of more than 95 percent of the vulture population.

---

* In India, vultures consume other sources of meat beyond cattle. A religious group that had traditionally exposed their dead to the elements rather than burying them had to stop the practice, as the birds that once quickly consumed the flesh were vanishing.

Necropsies conducted on recently dead vultures revealed that many (up to 85 percent) died from acute renal failure. Additional testing on tissue from the dead birds looked for the usual suspects associated with avian renal failure—cadmium, mercury, and infectious diseases including avian influenza, infectious bronchitis, and West Nile viruses—yet all proved insufficient to explain the dramatic die-off. Given that the primary food source for the vultures is cattle carcasses, a survey of veterinarians and veterinary pharmaceutical retailers turned up a potential chemical suspect, the non-steroidal anti-inflammatory drug diclofenac. When kidney tissue from the vultures was tested for the drug, the results were striking. All of the tested vultures (25 out of 25) that had died of renal failure—and none (0 out of 13) of those that died from other causes—contained residues of diclofenac in their livers. Toxicity testing, in which oral doses of the drug as well as tissue from livestock that had been administered diclofenac were administered to captive vultures, confirmed the acute toxicity of the drug.

Results from the research confirming the toxicity of diclofenac were striking for reasons that extend beyond India's borders. Residues of a pharmaceutical agent within the tissues of livestock were causing a population crash in a predatory bird, even when the drug was used responsibly. Unlike the situation that occurred in the 1960s and 1970s with bald eagles and DDT, there was no biomagnification of the chemical as it moved up the food chain; it was more simply a case of the legitimate use of a veterinary pharmaceutical leading to the decimation of a vitally important species within a long-standing and functional ecosystem.

The dying vultures illustrated a short and direct route between pharmaceutical exposure and harm to a wild animal. Kolpin's research, on the other hand, presents an exposure route from wastewater to aquatic organism that is more diffuse and occurs at lower concentrations. Is there evidence that exposures such as these represent a serious threat to aquatic wildlife such as fish?

One reason to think that PPCPs may be toxic at low concentrations comes from the relationship between the antidepressant fluoxetine (also known by the trade name Prozac) and fish. In Kolpin's geographic study of PPCPs in stream water, fluoxetine was detected at only one of the eighty-four analyzed sites. Furthermore, the estimates for in-stream concentration were low, in the 10 nanogram per liter range. More recent research from the United Kingdom supports the contention that in-stream concentrations are likely to range between 10 and 100 nanogram per liter (parts per trillion) wherever fluoxetine is detected. Given that the concentrations of fluoxetine are so low, can there really be a problem?

Since the publication of Kolpin's article, over thirty studies have been conducted to evaluate the effect of fluoxetine on fish. While the majority of the studies only demonstrated impacts on fish when exposed to concentrations between 30 and 100 micrograms per liter, some studies reported impacts on fish at much lower concentrations, rivaling those found by Kolpin in surface waters. The ideal concentration for fluoxetine in surface waters would be zero, as that would make the entire conversation regarding water-quality criteria moot. Unfortunately, as Kolpin and his colleagues have documented, we have already entered into an era in which human activities can be detected in surface waters across the country and around the globe.

One driving question may be related to the effects that these chemicals are having on humans. The chemicals identified by Kolpin and his colleagues are in such low concentrations, and the routes to human exposure so obtuse, that the risk becomes very difficult to assess. However, if we view aquatic animals as our modern-day canaries in a coal mine, is it prudent to ignore the impacts that these compounds are having in the environment? In the case of environmental occurrence of PPCPs, the enemy is clearly us, and the chemical signature of our PPCPs in the waterways of the world is not likely to disappear anytime soon. Pharmaceuticals and personal care products, in an abrupt and surprising fashion, have entered the environment as well as the conversation regarding modern toxicological contaminants.

Chapter 17

Determining Sex: Chemicals and Reproduction

Two roads diverged in a yellow wood,
And sorry I could not travel both
And be one traveler, long I stood
And looked down one as far as I could
To where it bent in the undergrowth . . .
— Robert Frost, "The Road Not Taken"

Recent studies that have focused on the response of fish to natural and anthropogenic environmental contamination have yielded some surprising results. For example, the recent seeding of an experimental lake in northwestern Ontario with ethynylestradiol, the primary synthetic estrogen found in human birth control pills, led to the collapse of the fathead minnow population that was living in the lake. In a similar vein, male fish collected downstream from wastewater treatment plants in the United Kingdom were found to experience differing levels of intersex, the condition in which mature ovarian follicles and mature spermatozoa occur in the same individual. More disturbingly, the intersex males were reproductively compromised, being able to breed with variable levels of success depending upon the severity of the intersex.

These studies bring up some interesting questions. First of all, how is it that the fish can be so reproductively plastic that low levels of contaminants in the water can so adversely impact their survival and reproductive performance? Second, why fish? Would the same thing have occurred eventually if the animals were mammals, such as otters or dolphins?

Chemical impacts on wildlife reproduction have been well documented, and in a manner similar to that delineated in *Silent Spring* by Rachael Carson, it was the large fish-eating birds of prey that first brought public attention to the problem. During the 1970s, from the coast of Florida to California to the Great Lakes, fish-eating bird colonies were declining in population. Unlike the situation described in *Silent Spring*, the birds were not dropping dead, but were simply showing no desire to mate. The choreographed movements of pair bonding and nest building were conspicuously absent. In some locales, the birds produced eggs that did not hatch or that, once hatched, revealed deformed chicks. In other locales, nests were being occupied by two females, with double the normal number of eggs.

One person who helped to piece the puzzle together was Dr. Theo Colburn, a scientist then working for the Conservation Foundation. She saw that chemicals such as ethynylestradiol, mentioned previously, were entering the birds' bloodstream and masquerading as cell signals, altering their reproductive physiology, morphology, and behavior in troubling ways. To understand her study of endocrine disruption, we first need a basic overview of reproductive biology, starting with the most familiar vertebrates, the birds and mammals.

The reproductive system of vertebrates produces one of two different outcomes, male or female, as the testis and the ovary arise from the same primordial tissue. In mammals and birds, sexual differentiation is genetic, and once the genetic sex of the individual has been determined, it remains comparatively stable. Gonadal differentiation generally proceeds down a single developmental path to yield fully differentiated testes or ovaries. In mammals, this outcome is determined by a gene known as SRY, localized on the short arm of the Y chromosome. Mammals with a Y chromosome, and thus with an SRY gene, develop into males.

The situation is very similar in birds, except that the sex with the homologous pair of chromosomes (ZZ) is a male, while the variant (ZW) is female. Recent research has indicated that a single gene, DMRT1, initiates the pattern of development that results in a male bird. Male birds in which DMRT1 activity has been knocked out are feminized.

To understand how a gene such as SRY or DMRT1 determines sex, it is necessary to examine how internal genitalia develop. Early on, the primordial gonad begins to mature without germ cells (eggs and sperm); these cells migrate into the gonad later in the process. The internal genitalia of the reproductive system (the vas deferens, epididymis, and seminal vesicle in males, and the oviduct, uterus, and upper part of the vagina in females) originate from the developing kidney ducts, and these primordial ducts have different fates in males than in females.

During development, a portion of the mesonephric kidney develops into two different ducts, the Wolffian and Müllerian ducts. The Wolffian duct, under the influence of testosterone, differentiates into the vas deferens, the epididymis, and the seminal vesicle. A second pair of embryonic ducts, the Müllerian ducts, develops alongside the Wolffian ducts. In males, the Müllerian ducts are suppressed due to the action of Müllerian-inhibiting substance (MIS), a hormone secreted by the testes. Maleness depends upon the secretion of testosterone from the testis, and in the absence of testosterone a male will develop a female phenotype. The SRY gene apparently activates the synthesis of MIS, which in turn assures that the Müllerian ducts will atrophy and that the mammal develops as a male.

In mammals, females are the result of the default reproductive developmental architecture. If the Müllerian ducts are allowed to develop, they will differentiate into the oviducts, the uterus, and the upper parts of the vagina, although the full development of a phenotypic female also requires estrogen. Furthermore, in females the Wolffian ducts, no longer augmented by testosterone, degenerate.

In mammals and birds, the developmental processes described above are largely hardwired: a gene on the sex chromosome determines the sex of the individual. However, there is more to reproductive function than genetic sex determination. An adult mammal's reproductive system was shaped by the steroids it encountered when developing as a fetus. The sensitivity of the system to steroids was inherent from the beginning; it was just a matter of looking for it in the right place. For mammals the right place to look was the placenta.

When an ovum implants into the uterine lining, a *placenta*—an intricate capillary network—begins to develop between the mother and the embryo. Capillaries from the mother and fetus anastomose, or mingle, closely together. While the mother's blood does not come into direct contact with the fetus's, oxygen, nutrients, and other chemicals, such as cell signaling steroids can be passed from the mother's system to the developing fetus.

Identical twins, which develop from the splitting of a single ovum, share this intimate connection with both their mother and each other. Since identical twins are always of the same sex, the steroids that they exchange are consistent. In humans, fraternal twins of different sexes are protected from exposure to the "wrong" steroids because they do not share a placenta, each twin having its own.

But in other large mammals, such as oxen and cattle, chemical crosstalk becomes a problem. When the twin fetuses are of mixed sex, one male and one female, the inevitable result is a reproductively sterile female, known as a *freemartin*. Freemartins and their male twins share the same placenta, and thus they share some of the same hormones that are responsible for sexual development. Müllerian-inhibiting substance, one of the critical hormones in male development, is passed via the placenta from the male twin to its sister, where it interrupts the normal development of the Müllerian ducts into the female reproductive structures. Consequently, the uterus of the female twin does not develop and the animal is sterile.

When two fetuses share a placenta, the chemical comingling of steroids is overt and powerful; the impact on the female fetus is devastating. In other mammals, where litters of pups are the norm, subtle differences in the uterine environment impact the developing fetus in surprising ways. For rodents and swine, the uterus is effectively a cylinder with the fetuses lined up in a row. Other than the individuals at either end of the uterus, each littermate is located in between two other individuals, and there are consequentially only three options; your immediate littermates can be either two females, two males, or a male and a female.

The sex of your immediate littermates makes a difference, and is an important source of reproductive variability, as the chemical environment created by littermates is significant. Female rodent pups developing between two male littermates have higher levels of fetal testosterone, are more sensitive to testosterone as adults, and produce more male offspring, relative to female pups developing between two males. These females are also more likely to mount other females, have larger home ranges, and be more aggressive. Likewise, males developing between two males are more aggressive, have greater home ranges, exhibit more-pronounced parental behaviors, and are more sensitive to testosterone than are males that develop between two females.

The studies with littermates clearly demonstrate both the overt and the more subtle effects that endogenous chemicals can have on fetal development. The chemicals impacting the fetus, just like the chemicals being released by the plastic culture tubes used in Ana Soto's study, as outlined in chapter 14, don't have to come from the mother or a sibling, nor do they even have to be natural. Instead, these compounds can come from external sources such as the effluent being released from many wastewater treatment plants, or the runoff from some agricultural landscapes.

The first evidence of this phenomenon came not from people, but livestock. It has been well documented for over fifty years that pregnant sheep grazing in pastures dominated by red clover (*Trifolium pretense*) can experience impaired fertility. Subsequent experiments have shown that red clover contains a mix of isoflavones—naturally occurring phytoestrogens that can cause reproductive failure in female livestock. While the impacts of clover consumption are reversible (by moving the ewes to different pastures), the isoflavones create their effect by masquerading as a natural, but in this case inappropriate, cell signal.

Another example builds on Arthur Herbst's epidemiological studies (discussed in chapter 13) detailing the adverse impacts of DES on women's reproductive systems. Diethylstilbestrol, while not a steroid, is an endocrine-disrupting compound. It triggers structural defects by altering pattern development through inappropriate cell signaling. Specifically, exposure to DES during early development leads to an incomplete atrophy of Wolffian ducts, and this oversight cascades through development, ultimately causing reproductive abnormalities. The DES story tragically illustrates that endocrine disrupting compounds can lead to permanent damage.

Humans view sexual differentiation from a mammalian perspective, and generally consider it to be a hardwired connection between genetics and morphology. Yet even in mammals, freemartins and littermates illustrate that genetics is linked to morphology via chemical cell signaling, and that signals received through the placenta can obfuscate the message sent from the genetic material to the developing sexual structures. Furthermore, while sexual differentiation in birds is subject to comparatively modest alteration, it is much more plastic and prone to alteration in the reptiles and amphibians, reaching its ultimate flexibility in fish. In the ectothermic vertebrates (particularly fish), genetic sexual differentiation runs the gamut from obligatory to nonexistent.

Genotypic sex determination occurs in reptiles in a variety of different forms. In snakes, the female is heterogametic (analogous to the XY chromosome pairing in male mammals), whereas either the male or the female in some species of turtles and lizards is heterogametic sex. Still other lizards have genotypic sex determination without heterogametic chromosomes. A similar situation occurs in amphibians, in that some species have heterogametic males (XY), whereas others have heterogametic females (ZW). In fact, the Japanese Wrinkled frog (*Rana rugosa*) contains populations with heterogametic males and others with heterogametic females, suggesting that the evolution of sexual differentiation in this species remains in progress. Interestingly, the breeding of these homogametic males and females (XX females bred together with ZZ males) leads to the development of XZ offspring, all of which are male.

There are other amphibian examples that illustrate the lack of fidelity between genetic sexual differentiation and morphological sexual differentiation. The African clawed frog, *Xenopus laevis*, is a female heterogametic species (ZW females, ZZ males). However, exposure of larval frogs to estradiol can lead to an incongruity in which a genetic male (ZZ) displays the full internal and external genitalia of a female. Importantly, when these incongruous (ZZ) females are paired with ZZ males, they produce viable offspring, all of which will be both genetically and morphologically male.

In some reptiles, and a small number of amphibians, sexual determination primarily depends upon temperature, not heterogamy. It has been postulated that thermal sex determination may not depend upon a master sex gene (such as MIS in mammals or DMRT1 in birds) but rather upon a "parliamentary system" of control in which changes in the thermal environment within the egg lead to wholesale changes in enzymatic activity across many of the proteins responsible for sex determination.

While all of the variants in the control of sexual differentiation discussed above are impressive, it is within the 24,000 different fish species that sexual differentiation reaches its most variable form. As with the reptiles, sexual differentiation can involve heterogametic males (ZW), heterogametic females (XY), or neither. Hermaphroditic species generally do not exhibit strong genetic sex-determination systems, and while some fish are gonochoristic, developing as one sex or another and maintaining that sex throughout their lives, other fish species spanning no fewer than twenty-five taxonomic families exhibit some level of natural hermaphroditism. Hermaphrodites can be sequential, starting out as either male first and changing later into females, or females first and ultimately changing into males, or can be synchronistic hermaphrodites with functional ovaries and testes occurring simultaneously in the same adult individual.

In gonochoristic species possessing either ovarian or testicular tissue, hermaphroditism and spontaneous sex reversal are rare. Yet despite the rarity, abnormal hermaphrodites or intersexes have been observed within field and laboratory groups of fish. Intersex and abnormal hermaphroditism can be induced in fish by exogenous chemicals. Furthermore, ample evidence exists that gonadal differentiation can be reversed in gonochoristic fishes. Exposures to an androgen or androgenic compound have led to regression of ovarian tissue and to the induction of testes that produce functional sperm. Likewise, examples exist in which females can be produced with estrogens even after the testicular development had begun.

The flexibility of sexual determination in nonmammalian vertebrate systems allows for much more dramatic changes within the reproductive system than those which occur in mammals or birds. This has been known by aquaculturists for decades, as there are management situations that require introducing sterile populations of animals into the wild. One way to ensure that these fish cannot breed is to produce single-sex populations by altering them chemically. Applying steroids during the appropriate stages of early development can create all-male or all-female populations. Alterations in sex ratio, though not as dramatic, are also found in fish living downstream from wastewater treatment plants and pulp and paper mills.

Developmentally, the reproductive system of vertebrates has two outcomes, male or female. While it is convenient to envision these two outcomes as dichotomous, the reality is not so simple. As discussed above, genetic males do not always develop into morphological ones, not even in mammals. Changes in the chemical environment in which fetal or larval animals develop are important, as they can disrupt the endogenous endocrine signals, causing alterations in reproductive development. Furthermore, the plasticity of the reproductive system is greatest in fishes, and the response to exogenous chemicals is far and away the most plastic in these animals. Whether it is reduced reproductive performance or wholesale collapse of a fish population, as per the experimental lake study in northwestern Ontario, these chemical signals can dramatically alter the populations and by doing so can cause dramatic impacts on the ecosystem.

Chapter 18

The Earliest Exposure: Transgenerational Toxicology

Children aren't coloring books. You don't get to fill them with your favorite colors.
— Khaled Hosseini

When mammals are exposed to chemicals during fetal development, the outcomes can last a lifetime. In some cases, the catastrophic impacts can be due to chemical malfeasance, as exogenous chemicals masquerade as cellular signals, and alter pattern formation within the fetus. The production of freemartins or the adverse impacts of diethylstilbestrol on the reproductive systems of developing females are classic examples of such impacts upon tissue organization. Other chemicals alter development by altering the heritable material (in other words, the genes and chromosomes), and under these conditions the impacts may extend beyond the mother and her developing offspring, impacting future generations well after the original chemical exposure has subsided. This chapter focuses on the developmental origins of adult disease, and more specifically on multigenerational and transgenerational toxicology.

For toxic chemicals, as Paracelsus stated, the dose makes the poison. However, when it comes to multigenerational toxicology, it is important to consider just who is being dosed. For non-pregnant mammals, both male and female, an exposure dose normally only directly impacts the tissues of the parent generation (also known as the F0). If a female is pregnant, as was discussed in the last chapter, the toxin can impact both the parent and the offspring (the F1 generation). Chemicals that target the genetic material are able to impact the F1 generation, even when the exposure is administered to a non-pregnant female or a male. In this case, the genetic material within the F1 generation is reached through the germ line (the sperm and ovum) within the testes and ovaries of the adult. A pregnant female is a special case, in that the exposure may affect not only the mother and fetus, but also the germ line within the fetus, and by doing so, impact the F2 generation. In this case, the exposure of the mother impacts the genetic material of her grandchildren.

The effects delineated above are known as *multigenerational* effects, as even the grandchildren of pregnant females have a direct, albeit limited, interaction with the parental exposure. As was discussed in the previous chapter, overt changes in pattern development can occur if the toxic chemical, such as diethylstilbestrol, alters cell signaling. But what if the exposure during development doesn't overtly alter tissue organization, but rather alters the susceptibility to disease later on in life? This hypothesis, known as the *developmental origins of disease*, is not new; as it was identified more than seventy-five years ago. In the earliest cases, the exposure was not due to a chemical, but rather was nutritional, with the most well-studied of the cases being the Dutch famine of 1944–45.

In 1944, the European component of World War II appeared to be reaching an endpoint. Allied forces landed on the beaches of Normandy on June 6, 1944, and by the end of the summer, Paris, Belgium, and the southern Netherlands had been liberated from the Germans. The movement of Allied troops stalled at the Rhine, and the northern Netherlands remained under German control as winter approached. In an effort to assist the Allied troops, the Dutch government-in-exile called a general rail strike to disrupt the flow of German troops and equipment into the region. In response, the Germans cut food and fuel imports into the occupied region of the country. By early November the canals had frozen over, the region was cut off, and the famine began.

The famine lasted until the Netherlands was liberated from Nazi rule in May 1945. During the famine, food consumption per capita was reduced from 1800 calories per day prior to the embargo to a low of 400 calories per day. During this time, women continued to have children despite their prolonged exposure to restricted rations, and despite the famine the medical records regarding these newborn infants were meticulously recorded. Long after the war, babies born at the Wilhelmina Hospital between November 1, 1943, and February 28, 1947, were eligible for inclusion in a Dutch famine birth-cohort study. Fifty years later, disease records for the adults who survived a fetal exposure to famine were matched with information collected during pregnancy, including the mother's weight and health throughout the pregnancy, as well as the size of the baby and the placenta upon birth. Famine exposure was divided into early, mid, and late gestation periods, depending upon the date of the child's birth relative to the famine's occurrence.

Comparisons between the health of the adults who were born during the famine and those born after were striking. Babies exposed to famine in late gestation were smaller at birth, and were found, as adults, to have impaired glucose tolerance relative to non-famine adults. The babies exposed to famine during early gestation were not smaller at birth, suggesting that the fetus underwent compensatory, or catch-up, growth during the latter parts of gestation, prior to birth. Nevertheless, the adults who developed from these children had a three-fold increase in coronary heart disease, a lipid profile in the blood more likely to deposit fat in the arteries (contributing to atherosclerosis), and a greater likelihood to be obese. Furthermore, the proportion of people within this group who self-reported being in poor health was statistically greater than that of the other groups.

The Dutch famine is a poignant example demonstrating an underlying association between fetal development and disease conditions in adults. Surprisingly, this result may actually be adaptive, in that the metabolic changes that occur during development may better prepare the offspring for a life in a stressful environment, such as one of food restriction. Had the babies that were exposed to famine in early gestation lived their entire lives under conditions of food restriction, the metabolic changes that occurred in them might actually have been adaptive to that environment. Problems arise, according to the developmental origins of adult disease theory, when there is a mismatch between the mother's environment (food restriction) and the environment that the offspring lives in (food abundance).

Results from the Dutch famine have led scientists to question how such developmental changes can occur, and in subsequent research a powerful mechanism for such changes has proven to be the epigenetic modification of the genome. *Epigenetics* is the study of stable alterations—alterations that can be passed from one generation to the next—in gene expression potential that arise during development. Epigenetic alterations, as opposed to genetic mutation, would occur "above" the genetic material; in other words, the alterations would not involve mutation or alteration of the DNA sequence itself.

To elucidate how these epigenetic modifications may occur, it is important to take a closer look at the genetic material. Chromatin, the collective material that becomes tightly wound into chromosomes during mitosis, is not merely the DNA strand, but rather a complex of macromolecules, including proteins known as histones, single-stranded RNA, and double-stranded DNA. The fundamental subunit of chromatin is the *nucleosome*, which is composed of eight histones, forming a core or spool which is encircled by a few wraps of the DNA string. This arrangement, while economical, necessitates that the DNA be unwound when it is read and transcribed into

RNA. Epigenetic modification of the chromatin alters the capacity of certain nucleosomes to unwind, thereby preventing gene transcription and translation.

Epigenetic modification can occur in at least three different ways: modification of the DNA itself, modification of the histones, or alteration of the noncoding RNA. One common epigenetic modification involves attaching methyl groups to the DNA. The attached molecules, known as "marks" or "tags," do not cause changes in the DNA strand itself, but rather allow for methyl groups to bind to it.

To understand where the methylation is most likely to occur within the genetic material, it is important to recall how DNA is structured. The DNA molecule is a double helix, a ladder-like macromolecule that spirals along its long axis. The ladder is made up of nucleotides that consist of a nitrogenous base, a five-carbon sugar, and at least one phosphate group. There are four nitrogenous bases, adenine, thymine, guanine, and cytosine, as base pairs for the rungs of the double helix ladder. The four nitrogenous bases always pair together such that adenine binds with thymine, while guanine binds with cytosine; therefore, the rungs of the ladder are composed of either A-T or C-G nitrogenous base pairs. Methylation tends to occur at C-G islands—regions within the DNA that have high proportion of C-G (rather than A-T) nucleotide linkages—and typically this binding causes a suppression of gene expression. A second form of epigenetic modification is known as histone posttranslational modification, in which one of a number of side groups (e.g., phosphates, acetyl groups, or methyl groups) attaches to the histones, thereby altering the accessibility of the DNA to transcription factors and in turn limiting expression. The third form of epigenetic modification involved the noncoding RNAs that are involved in either facilitating or inhibiting gene expression. It is noteworthy to recognize that these mechanisms of epigenetic modification do not occur in isolation from each other, but rather act in concert with each other to silence or promote gene transcription and translation into protein.

While the epigenetic events discussed are well documented, their importance comes to light when comparing the survivors of famine to control individuals. Individuals who were exposed to the Dutch famine during early gestation had, sixty years later, less DNA methylation of a specific gene (insulin-like growth factor 2) when compared to a same-sex sibling that was born under normal food conditions. Interestingly, when the comparison was made between siblings, one who had experienced normal food conditions and the other who had experienced famine at the end of gestation (as opposed to the period of early gestation), the relationship disappeared, and there was no difference in methylation. A subsequent study attempted to generalize this result by conducting the same comparison, but using global DNA methylation assays did not reveal any statistically significant hypomethylation, suggesting that the observed hypomethylation is a targeted rather than a generalized event.

The Dutch famine provides evidence regarding the role of epigenetic modification and adult disease. A second line of evidence arises from studies featuring identical twins. As discussed in the previous chapter, identical twins originate from a single fertilization event, and the single zygote splits into two genetically identical zygotes. At this point in development, the genetic material within the two individuals is the same, as are their epigenetic modifications. During the course of a human lifetime, epigenetic modification, including genetic methylation, builds up within an individual as a response to environmental effects such as chemical exposure, low-dose radiation, nutritional status, and behavioral cues. The consequence of the differences between the environment of the twins leads to "epigenetic drift," such that the level of DNA methylation and modification diverges between the two individuals. The incidence and timing of heritable age-related diseases, such as type-2 diabetes, Alzheimer's disease, and various types of cancer, including breast and prostate cancer, do not occur in direct concordance between identical twins, suggesting that the environment plays a role along with the genetics. The fact that epigenetic modification has been linked to a variety of diseases provides supportive evidence to the contention that epigenetic differences among individuals are associated with different disease conditions.

Chemicals and Epigenetic Modification

Recent research with toxic chemicals has found that endocrine-disrupting chemicals, such as diethylstilbestrol; genistien, the phytoestrogen derived from soy; and bisphenol A, a plasticizer used in the manufacture of polycarbonate plastics, all cause epigenetic effects. For example, when DES is administered during early neonatal development, the expression of a number of key developmental genes is upregulated due to epigenetic hypomethylation. Furthermore, these changes in gene expression are not transitory but rather are maintained well into adulthood.

When taken together, the multigenerational effects of endocrine-disrupting chemicals on pattern development become more complicated as alterations in cell signaling interact with epigenetic modification. An endocrine-disrupting compound such as DES can act through a steroid receptor and initiate a cascade of gene expression that will alter expression of steroid-responsive genes. Epigenetics can also play a role in this system, as activation of some of these steroid-responsive genes will depend upon their methylation state. As such, epigenetic modification also plays a role in gene expression well after, or downstream of, the action of a signal mimic as it inappropriately interacts with a steroid receptor. Endocrine-disrupting chemicals such as DES or BPA, therefore, can act to stimulate gene expression both at the receptor site and at sites farther downstream by altering cell signaling and by influencing DNA methylation.

Transgenerational Effects

Multigenerational effects due to exposure to toxic compounds occur in generations that were present, at least in the form of reproductive germ cells, during the parental exposure. Transgenerational effects, on the other hand, occur in animals long after the exposure had occurred. At first blush, it does not seem possible that an exposure occurring during pregnancy could possibly adversely impact subsequent generations beyond the granddaughters—that is, the F2 generation. However, recent research with rodents has demonstrated that multigenerational effects linger well past the F2 generation into the F3 generation and beyond.

The first recorded incidences of chemically induced transgenerational effects occurred after the F0 generation of rats were exposed to the pesticide vinclozin, a fungicide that is also an antiandrogen. Serendipitous breeding of the F1 offspring with each other found that the vast majority of males produced in the F2 generation carried a specific defect within the sperm (increased programmed cell death, or apoptosis). Surprisingly, this defect was maintained in subsequent generations, all the way into the F4 generation. Additional breeding studies found that outcrossed male offspring born to a naïve female (not from the exposed lineage) maintained the effect on subsequent generations of males, whereas an outcrossing of a naïve male with an exposed female did not, indicating that the transgenerational disease condition was primarily transmitted through the male germ line.

Since the early findings on vinclozin, the transgenerational effects have proven to be more involved than just sperm apoptosis, as a suite of adverse outcomes are being seen in subsequent generations. In addition, other studies have shown that other endocrine-disrupting chemicals, such as dioxin; methoxychlor, a pesticide mixture including permethrin; N,N-diethyl-m-toluamide (DEET); and hydrocarbons (jet fuel), have also been found to produce transgenerational effects that are often transmitted down one side (male or female) of the germ line.

How can it be possible that an exposure to an individual's great-grandmother could possibly impact the great-grandchild? Far from being science fiction, these effects, known as *transgenerational effects*, are well documented, and the mechanism underlying the effect may very well be the epigenetic response discussed above. To understand how this may happen, it is necessary to take another look at epigenetic modifications and how they are dealt with from one generation to the next.

When considered within the lifetime of individuals, such as the identical twins mentioned above, methylation can be viewed as a progression in which the genetic material is progressively more and more modified over time. This is not, however, the case from one generation to the next, as there are points in development in which the epigenetic slate is basically wiped clean. During embryonic development and gonadal sex determination, the germ cells undergo genome-wide demethylation, erasing methylation marks that accumulated within the parent.

An interesting point about the demethylation that occurs in early development is that right after it occurs, the epigenome begins to be methylated once again. This is absolutely necessary, as methylation is a fundamental method by which some genes are silenced. Furthermore, epigenetic modifications actually allow some genes (known as imprinted genes) to be expressed only by the parent that contributed them. Imprinted genes violate the normal rules of inheritance dictating that the genetic material from both parents be equally expressed. In other words, genes from one side of the lineage (often the maternal side) are silenced while the genes from the paternal side are expressed. This alternate expression may be adaptive in that the genes from the paternal side may be those that provide the developing mammalian fetus with resources from the mother, whereas the genes from the maternal side would be those that would more evenly distribute energetic resources between mother, the developing fetus, and whatever other siblings were currently undergoing gestation. As such, the process of imprinting is a epigenetic mechanism by which the father can ensure that his offspring are well taken care of during maternal gestation.

The scientific consensus at this time is that chemicals such as vinclozin exert their impact on subsequent generations in an imprinted gene-like fashion. After the methylation slate has been wiped clean, the epigenetic marks associated with these disease conditions, in the same manner as a benign imprinted gene, are quickly re-methylated and reestablished within the epigenome. In a manner similar to the masquerade of an endocrine-disrupting chemical as a cell signal, exposure to some toxic chemicals causes epigenetic changes that are maintained in the germ line for at least four generations. A normal and necessary biological process is being hijacked by an exogenous chemical, leading to prolonged adverse impacts.

Epigenetics truly challenges the ideas and definitions of classic toxicology. If a chemical causes a change in the genome that continues to create adverse impacts on populations that are generations removed from the original exposure, does the effect remain a toxicological question, or has it morphed into an issue related to development biology or molecular genetics? Furthermore, do efforts to regulate the use of chemicals such as vinclosin and methoxychlor necessitate that companies need to test the safety of chemicals for their effects out well beyond the third generation?

Chapter 19

Natural Toxins Revisited

Measles make you bumpy
And mumps'll make you lumpy
And chicken pox'll make you jump and twitch
A common cold'll fool ya
And whooping cough can cool ya
But poison ivy's gonna make you itch
— Jerry Leiber and Mike Stoller, "Poison Ivy" (as sung by The Coasters)

Chapter 8 focused on natural poisons and dealt with chemicals that were directly involved in the day-to-day arms race between species. A rattlesnake's venom, for example, has to neutralize a rodent so that the snake can have dinner. Toxins such as these have to be administered via a spine or a fang or a stinger. In other cases, the poison is not delivered, but rather lies within the animal's body, and exposure only occurs when a predator eats, or attempts to eat, the poisonous prey. The tissues of the blowfish, for example, are edible but contain tetrodotoxin, and extreme care in preparation is necessary lest the diner receive a potentially lethal dose of the neurotoxin.

In the cases above, there is no real secondhand environmental exposure, as the intended route of exposure is the only one that it truly viable. However, there are other examples of natural toxins that don't exactly fit this mold, and a review of those poisons illustrates a disconnect between the producer of the toxic molecule and the recipient of the toxic dose. In other words, these natural poisons can find themselves in the environment separated from their hosts, but still powerful with respect to toxic impact.

One of the best-known of the biologically mediated environmental toxins is urushiol, the oily compound produced by poison ivy, poison oak, and poison sumac. Urushiol produces an allergic reaction on contact with skin, but curiously the chemical is not secreted onto the outer surface of leaves or roots, but rather is internal to the plant. Gently rubbing up against an intact leaf will not cause a rash unless the plant has been mechanically damaged and urushiol has been released. Unfortunately, these plants are very fragile and even tiny holes in the leaves made by chewing insects will release the compound; therefore it is rare for the casual passerby to encounter an undamaged plant.

Once the urushiol is released from the plant tissue into the environment, its toxic potency, as well as its sustainable nature, become readily apparent. Urushiol is a fat-soluble compound and will not wash off with water alone. Furthermore, urushiol packs a significant wallop, as only a nanogram (a billionth of a gram) is necessary to cause a rash. Not only is urushiol highly potent, it is also incredibly persistent, as there are cases in which centuries-old urushiol has been found to retain its toxic potency. Plant parts—the leaves, stems, and roots—do not have to be alive to remain toxic, as the oil can remain active on any surface, including dead plants, for five years or longer. The oily nature of the compound, its high potency, and its persistence in the environment all act together to create a maddeningly annoying toxin. Secondhand contact (the horizontal transfer of the chemical from plant to pet fur to human skin, or from plant to clothing to human skin) is not only possible but likely, and it is not uncommon for the family pet to transfer the oil from the backyard to an owner's hands. Likewise, urushiol on the clothing of one individual can be administered to a second long after the initial environmental exposure has occurred. Furthermore, when the plant is mowed or trimmed, or when vegetation is burned, the chemical can disperse into the air, where it can continue to elicit toxic dermal effects. Inhalation of the aerosol is particularly dangerous, as the chemical can cause skin rashes as well as irritation and swelling within the tissues around the eyes and within the throat and lungs.

Plants are not the only organisms that produce such environmentally transmittable toxins. A number of different bacterial species produce *exotoxins*, which are chemicals that, once secreted, act at a site removed from the bacterial growth. Exotoxins are usually proteins that interact with host cells, producing a wide variety of responses, and most exotoxins act at tissue sites remote from the bacteria that produce them. Many of these exotoxins cause disease, and some well-known and historically devastating diseases, such as botulism, diphtheria, and tetanus, are caused not by the bacteria themselves but rather by the exotoxins that they secrete. A perfect example of this phenomenon are the neurotoxins produced by the bacteria *Clostridium botulinum*.

There are no fewer than seven different forms of botulism neurotoxin, with one of the forms, botulinum A, being the most toxic substance currently identified. Botulinum A binds to the presynapse membrane of nerve cells that secrete acetylcholine, preventing release of the neurotransmitter (see chapter 8). *C. botulinum* is an obligate anaerobe, meaning that it grows, multiplies, and produces neurotoxin only when found in an anaerobic environment. *C. botulinum* also produces spores that are both heat- and oxygen-resistant. While botulism poisoning, through improperly processed canned foods, is probably the way that most people are familiar with the neurotoxin, *C. botulinum* also has an environmental life cycle that relies upon the exotoxin for its completion.

Spores of *C. botulinum* are fairly ubiquitous in the environment in the sediments and waters found in many waterways. One condition that may favor the production of *C. botulinum* includes stagnant, warm water with a proliferation of filamentous mats of blue-green algae. These mats ultimately die and begin to decompose, producing small islands of anaerobic organic material in which *C. botulinum* can thrive. Once the bacteria begin to grow, they inevitably produce botulinum neurotoxins. Ducks or other waterfowl may succumb to the neurotoxins, and, once dead and decomposing, the waterfowl can become incubators for *C. botulinum*, which immediately begin to secrete more of the botulinum toxin. Maggots and other scavenging insects that feed on the decaying flesh are immune to the neurotoxin; however, they themselves become vectors for the bacteria and the neurotoxin. Waterfowl that eat the insects may die from exposure to the neurotoxin, decompose, and then themselves become incubators for the proliferation of the botulinum bacteria. The cycle (known as the carcass-maggot cycle) is self-perpetuating and can lead to outbreaks of avian botulism that can ravage waterfowl.

In the cases above regarding urushiol and botulinum neurotoxin, the toxin is well defined and its release has a well-defined target and goal. However, in other cases the toxin is less well defined and its occurrence may be more accidental. As an example, consider prions. *Prions* are proteins that can be folded in different ways, including a normally configured prion protein that is harmless and an infectious form that causes disease. The proteins have the same amino acid sequence; however, their three-dimensional shape (the way that the protein folds together) is different. The infectious prion has the ability to convert normal prion proteins into infectious ones, and as such the prions are self-replicating.

As more infectious prions are created, they aggregate together into fibrils, and it is these fibrils that appear to induce disease. Prions are the cause of a number of degenerative brain diseases, including bovine spongiform encephalopathy or "mad cow" disease in cattle, scrapie in sheep and goats, chronic wasting disease in the wild cervids (the deer, moose, and elk), and Creutzfeldt-Jakob disease in humans.

A prion is an *infectious agent*—a compound that falls somewhere in between an infectious disease, such as disease-causing bacteria or a virus, and a toxic compound. *Infectious diseases* are characterized by biological systems that contain genetic material in the form of RNA or DNA and that are initiated by viruses, bacteria, or fungi. Infectious diseases are also components of living systems, and once removed from the infectious host they are governed by biological constraints. Many virulent viruses and bacteria do not survive well outside of the host and have a very attenuated survival time on the skin or on hard, dry surfaces.

In contrast, an infectious prion replicates without the aid of DNA or RNA. Furthermore, its horizontal transmission from one individual of a species to another can occur either biologically or environmentally. Biological transmission in cervids can occur through the comingling of shared fluids. Cervids tend to be herding animals with elaborate social structures. Contact among individuals through saliva, urine, feces, or birthing matter is not only possible but likely as a means of chemical communication among individuals. As such, a prion infection within one individual may very likely spread to others within its herd.

Environmental transmission has also been shown to occur. In one example, sheep introduced into a field in which scrapie-infected sheep grazed sixteen years earlier also became infected with scrapie, suggesting a very long residence time for the infectious agent. Prions have been found to bind to certain types of soil particles, and even in the bound state remain bioavailable to animals that graze on these pastures. Unlike biological organisms, the prion molecule in the environment acts as a chemical. Similarly to urushiol from poison ivy or botulinum toxin from *Clostridium botulinum*, it can persist for a long time in the environment, independent of the biological organism that produced it in the first place.

With respect to human illness, Creutzfeldt-Jakob disease is a rare disorder that affects only one or two individuals per million annually. Of these cases, the vast majority (85 percent) occur not through horizontal transfer from animals or the environment, but rather through spontaneous conversion of the noninfectious form, found within all people, into the infectious one. The misalignment of the prion proteins leading to the production of the infectious prion most commonly affects people over sixty-five and is usually fatal within six months from the onset of symptoms.

A malady called variant Creutzfeldt-Jakob disease (vCJD), a variant of mad cow disease, is believed to be caused by consumption of beef products contaminated with brain or spinal cord from cattle infected with mad cow disease. While it remains a rare event (only four deaths in the United States have been attributed to it), the ability of mad cow disease to become infectious from cattle to humans has generated a great deal of concern regarding the proper disposal of infected cattle. Rendering—the process of converting unused organic material remaining after animal butchering into usable animal products—can deliver prions either into portions of the products themselves or into the wastewater stream. Likewise, complete carcass burial can present an avenue by which prions can enter the environment, where they may ultimately contaminate soils or water.

In most cases in nature, a toxin, naturally produced by an organism, comes at a cost to its producer, and as such, toxic chemicals that do not convey a distinct survival advantage to the producer represent an energetic cost without any fitness benefit. Under these circumstances, the toxic chemical is not likely to remain in production over generational periods of time, as the costs can far outweigh the benefits. Prions turn this idea on its head, as the prion is produced accidentally, not deliberately. Furthermore, a prion is a self-replicating entity; therefore the concept of a dose–response relationship is limited in its utility, as the concentration of prions within the neural tissue of an infected animal is likely to increase over time even without successive inoculations from the environment. At the same time, however, a prion does not behave in the environment like other natural organic chemicals, as they are persistent, and environmental exposure to bioactive prions is a very real prospect. Prions represent a special class of compound that is neither alive nor merely a toxic chemical, but rather shares properties of both.

Chapter 20

Chemical Resistance

If there's one thing the history of evolution has taught us, it's that life will not be contained. Life breaks free, it expands to new territories, and crashes through barriers painfully, maybe even dangerously, but, uh, well, there it is.
— Michael Crichton, from the screenplay for the movie *Jurassic Park*

Modern toxicology has taken the field in directions far beyond what Paracelsus could have ever imagined. Epigenetics challenges the idea that direct exposure is needed to cause harm, while biologic molecules such as prions show that toxicity cannot always be described by the classic dose–response relationship. How can the dose of a chemical be ascertained if the chemical is self-replicating? Another phenomenon on the fringe of toxicology involves the biological response to chemical toxicity, or resistance. Our widespread use of chemicals to combat pests has, somewhat paradoxically, provoked those pests to develop resistance. When taken to its endpoint, the development of resistance itself can be considered as a self-replicating "pollutant." But before that discussion, it may be instructive to consider how resistant populations of pest organisms develop.

Chemical Resistance and Natural Selection

Pesticides, and their functional cousins, antibiotics, represent chemical weapons leveled at species that we want to control. Although we have waged chemical warfare on insects, weeds, and other unwelcome species for only a relatively short period of time in evolutionary terms, our pests have still been able to develop resistance—and surprisingly quickly. Resistance to these chemicals is a product of natural selection, yet the concept is so deceptively simple that it may appear an inadequate explanation for the rapid development of defensive mechanisms. In fact, natural selection is a powerful tool in building chemical resistance.

The concept of natural selection begins with the premise that many species produce more offspring at any given point than will be able to survive and reproduce. As anyone with a sibling knows well, there can be substantial genetic variability among offspring. While the vast majority of genetic differences do not affect an individual's overall fitness, a few traits can be influential.* Furthermore, individuals with greater overall fitness generally contribute more offspring to the next generation than those individuals with lower overall fitness. In this manner, genes that confer substantial increases in fitness will proliferate through the population over time.

Individuals within a species must overcome numerous challenges to survive and reproduce, and chemical exposure is simply one of those challenges. Some of the exposed individuals may succumb to the toxin, whereas others may be able to resist. Individuals that remediate against the toxic impacts of chemical pesticides or antibiotics, through biotransformation or other techniques, are more likely to survive and pass their genes along to the next generation.

Species have developed a number of strategies to try to ensure that some of their offspring, and by extension some of their genetic material, survive and populate the next generation. One strategy is

* The term *fitness*, as used by geneticists, denotes the ability of an organism to survive and, more importantly, to reproduce.

the "big-few" gambit, in which the parents invest a lot of energy into their offspring, but by doing so they limit the number of offspring that they can produce. In this scenario the offspring are often long-lived, and the species has slow generation times. Certainly humans fall into this category, as do many familiar mammals, including domesticated animals such as cows, horses, dogs, and cats, as well as wild animals such as whales, deer, lions, tigers, and bears.

An alternative strategy is the "small-plenty" gambit, in which a large number of offspring are produced, but the investment into each individual offspring is relatively small. The idea is that, since the majority of offspring may not make it to reproductive maturity, it is prudent to invest as little in each individual as possible, and instead use that energy to produce as many offspring as possible. Consider the number of dandelion seeds produced per flower, or the number of mice that can be produced by a single pair over the course of a year. Species that adopt this reproductive, or ecological, strategy all share a number of characteristics, including the capacity to respond rapidly to disturbed environments.

Differences in reproductive strategy are important, as they provide a tip-off regarding a species' capacity to develop chemical resistance. All other factors considered, the more rapidly a species can proceed through reproductive generations, the more rapidly the species will evolve. The faster the generation time, or the more that a species has invested into the "small-plenty" gambit, the more likely it is that individuals from that species will develop resistance in a timely fashion. Unfortunately, the species that are the targets of chemical control—for example, rodents, mosquitos, aphids, corn borers, weevils, bacteria—are very often the species that have the capacity to develop chemical resistance the most quickly.

Insects and Pesticides

Agriculture, one of the stalwarts of human endeavor, is a cycle of cultivation and growth that is completed when humans eat what they have harvested. Yet other nonhuman consumers also have a taste for our food crops. From 10 to 40 percent of the plant material that we grow for food is lost to pests before the food ever gets to our dinner table. Fungi in the form of rusts and mildews take their toll on food crops, as do rodents; but the pests that probably generate the greatest agricultural ire are insects.

Insects have been challenged with chemical pesticides for centuries, but over the course of the last 100 years innovations in the development of insecticides have placed humans in a direct evolutionary arms race with many insect species. And, in that battle, it would be hard to conclude that the insects are losing, for during the last century, over 500 insect species have generated resistance to one or more pesticides.

Insect resistance to pesticides is partially caused by how we apply the pesticides. Consider a naïve field where no pesticides have been sprayed. The initial application of pesticide increases crop yield by killing a fair proportion of pesticide-sensitive insects. As the amount of pesticide increases, more insects are killed and crop yield continues to increase. The relationship is not linear, however, as there comes a point of diminishing return in which additional pesticide application only marginally increases yield. At this point, the economically astute farmer will stop applying the chemical, as the cost no longer realizes a consistent benefit in crop production.

Given that the susceptible individuals were killed with the pesticide, the survivors would be, by logical extension, only those that carry the genetic material for resistance. If all of the susceptible individuals within a geographic region are removed from the gene pool, the more resistant populations will breed with each other and proliferate. This is one reason why natural buffer strips as well as other uncultivated landscapes are so important in agricultural regions, as they give the susceptible pests a place to breed so that their genes are not removed from the insect population. Paradoxically, then, an effective method of pest control involves ensuring that genes susceptible to chemical control do not vanish from the overall gene pool. If the susceptible genes have been removed, then only pesticide-resistant genes will remain, and the whole population, much to the farmer's dismay, will develop pesticide resistance.

Pesticides follow the rules of absorption as described in chapters 4 and 5 of this book. The receptor site, loosely defined, is what is in contest, with the toxic molecules moving to bind to it, while the cellular defenses attempt to prevent that from happening. There are at least two primary mechanisms by which that binding can be thwarted, and the development of pesticide resistance in insects has examples of both. One strategy is to alter the receptor site so that the toxic compound can no longer bind to it. The second strategy is to deploy proteins within the cell that either convert the pesticide into a relatively nontoxic metabolite or reduce the sensitivity of the target site for the pesticide.

Regardless of the mechanism by which selection occurs, the development of resistance continues to reveal unexpected outcomes. For example, one expectation is that the alterations to the physiology and molecular biology of insects that makes them resistant should be ephemeral, as the selective benefit is only realized when the confrontation with pesticides is ongoing. While this is often the case, it is not always so. Specimens of Australian sheep blowfly (*Lucilla cuprina*) carried resistance-associated genes for malathion but not diazinon (two organophosphate insecticides) even among flies collected prior to the use of the pesticides. While these mutations may provide other selective advantages to the fly, the continuation of the resistant mutation in the lineage does not support the idea that pesticide resistance carries a cost to the fly in absence of the pesticide.

Antibiotics and Bacteria

An antibiotic is a drug that destroys microorganisms or inhibits their growth, while being nonlethal to humans or other animals. Originally, a distinction was made between the antibacterial drugs that were synthesized in a laboratory (such as the first commercially available antibiotics, the sulfa drugs), and antibiotics, which were produced by living organisms (such as penicillin). Today most antibiotics are derivatives of naturally occurring compounds that have been purified and are now synthesized; therefore the dichotomy between naturally produced and chemically synthesized antibiotics has been blurred.

Antibiotics have revolutionized medicine, as many infectious diseases have been largely controlled by their use. Consider the fact that during the Civil War over 70 fatalities per 1,000 troops occurred due to infectious disease. Approximately eighty years later during World War II, the fatality rate due to infectious disease had been reduced to less than 1 per 1,000. Partial credit for this profound increase in survivorship was due to the widespread use of penicillin as well as other antibiotics.

The history of penicillin is an interesting one. Its antibacterial properties were first discovered accidentally by Alexander Fleming in 1928.* Penicillin was treated as a curiosity until 1941, when research on its production coalesced through a network of laboratories in the United States. Within five years, penicillin production had developed from a crude, low-yield method to one making batches of penicillin via fermentation with industrialized mass-production techniques. Prior to 1941, laboratory scientists could only produce small quantities of crude penicillin; toward the end of World War II production had mushroomed to 4 million sterile packages per month.

Even during the very early days of the widespread use of antibiotics, microbial resistance was never very far behind. Long-term use (ten days or more) of a single antibiotic will select for bacteria not only resistant to that specific antibiotic but to others as well. Even without long-term use of antibiotics, resistance within hospital

* Fleming had gone on vacation to his country house, and upon his return to the laboratory he was cleaning petri plates that had become contaminated with fungi while he was away. A former laboratory assistant stopped by to talk with him, and he slowed down his normally brisk routine of disinfecting the plates for reuse. Fleming then noticed that some of the fungi that had colonized the plates had also inhibited the growth of the bacteria he had earlier inoculated onto them.

settings will develop over relatively short periods of time. For example, sulfa-drug-resistant bacteria began to appear in military hospitals in the 1930s, while penicillin-resistant *Staphylococcus*, the bacteria responsible for staph infections, became evident in London civilian hospitals in the 1940s. Indeed, bacterial resistance was one of the driving forces pushing the development of new antibiotics. Methicillin was developed in 1959 in direct response to bacterial resistance to penicillin. Vancomycin, an antibiotic with potentially toxic side effects, reached the market in 1958, but its use surged during the 1980s in response to the emergence of methicillin-resistant *Staphylococcus aureus*, and penicillin-resistant *Streptococcus pneumoniae*. The speed at which bacterial resistance can occur is astonishing. For example, when erythromycin was introduced in the early 1950s as an alternative to penicillin, it had to be withdrawn in less than a year due to a dramatic development of resistance in *Staphylococcus aureus.*

The capacity for bacteria to develop antibacterial resistance does not end at the door of the hospital or local health clinic, for the use of antibiotics outside of a clinical setting is booming. The exact amount of antimicrobials used to treat livestock or to enhance their growth is unknown, as national statistics are available for only a few countries worldwide. However, the World Health Organization estimates that at least half of the antibiotics produced worldwide is used in livestock, not humans. Given the state of industrial agriculture, in which thousands of animals are confined in small areas, adding antibiotics to feed can quickly provoke antibacterial resistance within the gut bacteria of the livestock as well as in the animals' waste stream. As an example, the time span between administration of tetracycline to chickens maintained in confined enclosures and the excretion from the chickens of coliform bacteria that had developed multi-antibiotic-drug resistance was just a matter of a few weeks.

The relationship between an antibiotic and the development of resistance depends upon the antibiotic density—in other words, the amount of it found occurring in a given geographical region. As previously noted, hospitals top the list, and the speed at which antibacterial resistance can develop in such environments is mind-boggling.

Antibiotics tend to be resistant to biotransformation, and the chemicals can persist long after they are excreted from the human or animal that they were originally administered to. Consequently, antibiotics are found in wastewater treatment plants, and also in areas where sewage sludge and animal manure has been spread.

Bacterial Plasmid as Pollutant?

When exposed to antibiotics, bacteria, like insects, can develop resistance by random genetic mutation. One example is the bacterium responsible for tuberculosis (*Mycobacterium tuberculosis*), which has developed extremely drug-resistant and totally drug-resistant strains exclusively by spontaneous mutation. However, unlike insects, bacteria also undergo a process known as horizontal gene transfer in which pieces of new genetic material are incorporated into other bacteria, which can then reproduce asexually via cell division. By taking advantage of horizontal gene transfer, bacteria have a much greater capacity for recombining their genetic material than do insects or any other multicellular organisms. Relative to the development of antibiotic resistance, the most common form of horizontal gene transfer is through the transfer of *plasmids*, small circular strands of DNA. DNA thus acquired from the environment may recombine with the host's DNA or, once inside the cell, may remain as a functional plasmid.

When certain bacteria begin to develop antibiotic resistance, a series of events can build upon the spontaneous mutation, ultimately leading up to multidrug resistance. Under chronic exposure, antibiotic-resistant genes that were once in the chromosomes of environmental bacteria are now packaged into plasmids that can be transferred to human pathogens. Similar to the self-replicating proteins known as prions (chapter 19), plasmids appear to be the most pervasive environmental pollutant associated with antibiotic-resistant bacteria.* The plasmids are self-replicating (through

* This does not mean to imply that the antibiotics themselves cannot be environmental pollutants. Many antibiotics, such as penicillin, arise from naturally occurring compounds and biodegrade in the environment relatively quickly. Furthermore, given the lack of self-

incorporation into competent bacteria), capable of traveling great distances, and are resistant to removal from the bacterial population even when no antibacterial agents remain in the environment.

As with insects, the origin of antibiotic-resistant genes in the environment is not directly associated with exposure to antibiotics. The only bacterial collection that exists from the pre-antibiotic era (prior to 1954) is the Murray collection of enterobacteria. Antibiotic resistance in these samples was negligible, supporting the fact that resistance pre-dates antibiotic occurrence, but also supporting that its occurrence was rare. Antibiotic resistance has also been found in bacteria harvested from remote locations, areas where the presence of antibiotics seems unlikely. For example, sediments harvested at two different field sites, from between 170 and 259 meters below the land surface, were found to contain over 150 strains of bacteria, with 90 percent of them being resistant to at least one antibiotic.

The preponderance of antibiotic resistance and antibiotic-resistant genes directly associated with humans reveals a somewhat different story than the recovery of antibiotic resistance in environmental samples. Inferences were made by evaluating the presence of antibacterial resistance in bacteria from human populations living in very remote environments, such as remote villages in Nepal and Bolivia. Even when these villages were only accessible by foot after an hours-long trek, the commensal enterobacteria associated with the local villagers still showed resistance for older antibiotics, including tetracycline, penicillin, and ampicillin, among others. More importantly, despite the remoteness of each village, the patterns of resistance were most like those found in the nearest urban area, regardless of its distance from the village. The simplest explanation is that occasional contact between villagers and the closest city-dwellers allowed for the dissemination of resistant genes and bacteria from the more urban group to the more remote.

---

replication, their concentration in the environment is low and their capacity to migrate long distances, particularly without the capacity to self-replicate, is certainly less than that for a self-replicating plasmid.

As with prions, the fate and transport of plasmids containing antibiotic-resistant genes is a significant concern regarding the spread of antibiotic resistance. Anthropogenic chemicals (antibiotics) have led to the proliferation of a biological entity (plasmids) that is environmentally transported (either in bacteria or in isolation within the environment) across wide distances to distant regions where they can elicit their impact. Of course, the fear is that these antibiotic-resistant genes may be transferred into complementary pathogenic microbes, creating resistance in bacteria far-removed from the antibiotics that they have developed resistance to. The antibiotic-resistance-carrying plasmid is neither a chemical nor an infectious agent, such as a prion. Perhaps more importantly, the antibiotic-resistance-carrying plasmid is another example of how far toxic compounds have come from those of Paracelsus and his elementary dose–response relationships.

Afterword

Toxicology and Beyond

In the places I go there are things that I see
That I never could spell if I stopped with a Z.
I'm telling you this 'cause you're one of my friends.
My alphabet starts where your alphabet ends.
— Dr. Seuss, *On Beyond Zebra!*

When researching fields of study across the toxicological disciplines, I have often been left wondering, *Why didn't someone tell me that before?* For example, it was not immediately obvious to me that transporter proteins and metallothioneins cannot readily be distinguished from metals from the same family, or column within the periodic table. I had to come to that realization over time. Likewise, the fundamental role that lipophilicity and biotransformation play in toxicology was not readily apparent but became evident only after I had been wrestling with these concepts. With this book, I want to bring those ideas to the forefront, not just to scientists and students, but to the interested layman as well.

My hope is that the reader now understands the core precepts of toxicology, as well as some of the most pressing challenges created by chemicals today. Yet I am mindful that this is a complex field that is expanding rapidly with new chemicals and more-subtle toxic responses. As in any decent scientific inquiry, many of the questions I have sought to address tend to fracture into an array of sub-questions.

For example, consider the chapter on particles. The smallest particles that the chapter focuses on are PM2.5—that is, particles that are 2.5 microns and less in size. Of course this is an oversimplification, as particles smaller than 2.5 microns can be further assigned to size classes that include the nanoparticles, those particles that are between 1 and 100 nanometers in size. While much attention has recently been given to manufactured nanoparticles such as those in some sunscreens, it is important to recognize that natural particles fall into this size range as well, including ultrafine particles in the atmosphere or colloidal material in the water. This material falls into a netherworld, in that such particles share characteristics with atoms and small chemical compounds and also with larger particles in the micron range. The relatively recent capacity to deliberately synthesize nanoparticles in the form of fullerines or nanotubules may revolutionize drug delivery, but this development is also likely to come with its own share of toxicological problems.

In another example, consider chapter 16 on pharmaceuticals and personal care products. Even though I mentioned that chemical mixtures of anthropogenic chemicals are commonplace in the environment, this book, like most other toxicology books, generally focuses on single compounds and the impacts that these compounds have on organisms and humans. But even a cursory understanding of toxicology reveals that mixtures are important and that they change impacts, sometimes in unexpected ways. There is a large body of literature suggesting that the combination of alcoholism with chronic smoking is much worse than either smoking or alcoholism alone. Likewise, the mixture of alcohol with pharmaceuticals, such as acetaminophen, influences the metabolic pathways that are responsible for the biotransformation of pharmaceuticals, such that the end result is the buildup of toxic metabolic intermediate compounds. While the mix of chemicals in the environment may be at such low levels as to be relatively benign, it is probably naïve to think that the combination of low concentration and environmental biotransformation is going to obviate all the problems associated with these mixtures.

And finally there is the 800-pound gorilla in the room: global climate change. Climate change is a toxic enabler, exacerbating phenomena associated with toxic compounds. In many areas worldwide, there will be greater oscillation in storm events such that the time intervals between events (floods) and the severity of the events may both increase simultaneously. Whether toxic compounds enter waterways from non-pervious surfaces (e.g., roads, rooftops), or from the solid sludge fields associated with wastewater treatment plants, or from agricultural row crops, the dynamics of the transport of these contaminants in the environment are likely to change, depending upon the frequency and intensity of rainstorm events. Climate change is also likely to affect processes of global distillation. As theses chemicals migrate back and forth from the ocean into the atmosphere, what role will changing water temperature play in that dynamic? It is not really clear, but the main point is that climate change will dramatically influence the fate and transport of toxic chemicals in the environment, perhaps in ways that now are not fully understood or anticipated.

"Not fully understood or anticipated" is a common phrase in toxicology, particularly as it pertains to the toxic compounds—pollutants—in the environment. This is a fast-changing field, with new studies and discoveries surfacing all the time. One thing is certain: humans do not stand apart from the chemical environment in which they live. When disposing of unwanted chemicals, either pharmaceuticals used in clinical settings or industrial chemical by-products, there really is no place on Earth that is "away."

As we continue to study the effects of far-flung chemicals, new questions will undoubtedly emerge. These questions will require engagement not only from toxicologists but from professionals in a variety of fields, from ecology to public policy. They should also interest anyone concerned with their own health or that of future generations. If this book piques readers' interest, or even better, prompts them to ask their own questions, then it has achieved its primary goal.

REFERENCES

Books

Carson, Rachel. *Silent Spring*. New York: Houghton Mifflin Company, 1961.

Colborn, T., D. Dumanoski, and J. P. Myers. *Our Stolen Future: Are We Threatening Our Fertility, Intelligence, and Survival? A Scientific Detective Story*. New York: Penguin Books, 1997.

Emsley, J. *The Elements of Murder*. Oxford, UK: Oxford University Press, 2005.

Kean, S. *The Disappearing Spoon: And Other True Tales of Madness, Love, and the History of the World from the Periodic Table of the Elements*. New York: Hachette Book Group, 2010.

Klaassen, C. D. *Toxicology: The Basic Science of Poisons*. 8th ed. New York: McGraw-Hill, 2013.

Mukherjee, S. *The Emperor of All Maladies*. New York: Simon and Schuster, 2010.

Randall, D., W. Burggren, and K. French. *Eckert Animal Physiology: Mechanisms and Adaptations*. New York: W. H. Freeman, 2001.

Stelljes, M. E. *Toxicology for Non-Toxicologists*. Rockville, MD: ABS Group, 2000.

Weston, A., and C. C. Harris. "Chemical Carcinogenesis." In *Cancer Medicine*, 8th ed., edited by W. K. Hong, R. C. Bast Jr., W. N. Hait, D. W. Kufe, R. E. Pollock, R. R. Weichselbaum, J. F. Holland, and E. Frei III. Shelton, CT: People's Medical Publishing House, 2009.

Papers

Chapter 1. The Dose Makes the Poison

Dorne, J. L., and A. G. Renwick. "The Refinement of Uncertainty/Safety Factors in Risk Assessment by the Incorporation of Data on Toxicokinetic Variability in Humans. *Toxicological Sciences* 86 (2005): 20–26.

Jager, T. "Bad Habits Die Hard: The NOEC's Persistence Reflects Poorly on Ecotoxicology." *Environmental Toxicology and Chemistry* 31 (2012): 228–29.

Landis, W. G., and Chapman, P. M. "Well Past Time to Stop Using NOELs and LOELs." *Integrated Environmental Assessment and Management* 7 (2011): vi–viii.

Chapter 3. The Human Animal

Adkins, R. M., E. L. Gelke, D. Rowe, and R. L. Honeycutt. "Molecular Phylogeny and Divergence Time Estimates for Major Rodent Groups: Evidence for Multiple Genes." *Molecular Biology and Evolution* 18 (2001): 777–91.

Andrews, P. L. R. "Laboratory Invertebrates: Only Spineless, or Spineless and Painless?" *ILAR Journal* 52 (2011): 121–25.

Berman, H. M., L. F. Ten Eyck, D. S. Goodsell, N. M. Haste, A. Kornev, and S. S. Taylor. "The cAMP Binding Domain: An Ancient Signaling Module." *Proceedings of the National Academy of Sciences USA* 102 (2005): 45–50.

Beyer, L. A., B. D. Beck, and T. A. Lewandowski. "Historical Perspective on the Use of Animal Bioassays to Predict Carcinogenicity: Evolution in Design and Recognition of Utility." *Critical Reviews in Toxicology* 41 (2011): 321–28.

Cohen, S. M., L. L. Arnold, M. Cano, M. Ito, E. M. Garland, and R. A. Shaw. "Calcium Phosphate-Containing Precipitate and the Carcinogenicity of Sodium Salts in Rats." *Carcinogenesis* 21 (2000): 783–92.

Gold, L. S., T. H. Slone, N. B. Manley, and L. Bernstein. "Target Organs in Chronic Bioassays of 533 Chemical Carcinogens." *Environmental Health Perspectives* 93 (1991): 233–46.

Gold, L. S., L. Bernstein, R. Magaw, and T. H. Slone. "Interspecies Extrapolation in Carcinogenesis: Prediction between Rats and Mice." *Environmental Health Perspectives* 81 (1989): 211–219.

Harvey-Clark, C. "IACUC Challenges in Invertebrate Research." *ILAR Journal* 52 (2011): 213–20.

Hurst, L. D., and N. G. Smith. "Do Essential Genes Evolve Slowly?" *Current Biology* 9 (1999): 747–50.

Jordan, I. K., I. B. Rogozin, Y. I. Wolf, and E. V. Koonin. "Essential Genes Are More Evolutionarily Conserved than Are Nonessential Genes in Bacteria." *Genome Research* 12 (2002): 962–68.

Kawakami, T., R. Ishimura, K. Nohara, K. Takeda, C. Tohyama, and S. Ohsako. "Differential Susceptibilities of Holtzman and Sprague-Dawley Rats to Fetal Death and Placental Dysfunction Induced by 2,3,7,8-teterachlorodibenzo-p-dioxin (TCDD) despite the Identical Primary Structure of the Aryl Hydrocarbon Receptor." *Toxicology and Applied Pharmacology* 212 (2006): 224–36.

Mather, J. A. "Philosophical Background of Attitudes toward the Treatment of Invertebrates." *ILAR Journal* 52 (2011): 205–12.

Putz, O., C. B. Schwartz, S. Kim, G. A. LeBlanc, R. L. Cooper, and G. S. Prins. "Neonatal Low- and High-Dose Exposure to Estradiol Benzoate in the Male Rat: I. Effects on the Prostate Gland." *Biology of Reproduction* 65 (2001): 1496–505.

Putz, O., C. B. Schwartz, G. A. LeBlanc, R. L. Cooper, and G. S. Prins. "Neonatal Low- and High-Dose Exposure to Estradiol Benzoate in the Male Rat: II. Effects on Male Puberty and the Reproductive Tract." *Biology of Reproduction* 65 (2001): 1506–17.

Rand-Weaver, M., L. Margiotta-Casaluci, A. Patel, G. H. Panter, S. F. Owen, and J. P. Sumpter. "The Read-Across Hypothesis and Environmental Risk Assessment of Pharmaceuticals." *Environmental Science and Technology* 47 (2013): 11384–95.

Rinaldi, J., J. Wu, J. Yang, C. Y. Ralston, B. Sankaran, S. Moreno, and S. S. Taylor. "Structure of Yeast Regulatory Subunit: A Glimpse into the Evolution of PKA Signaling." *Structure* 18 (2010): 1471–82.

Rothen, D., S. Baran, and M. Perret-Gentil. "Are Zebrafish the New Mice?" *ALN magazine* (2014). http://www.alnmag.com/articles/2014/06/are-zebrafish-new-mice.

Vargo-Gogola, T., and J. M. Rosen. "Modeling Breast Cancer: One Size Does Not Fit All." *Nature Reviews Cancer* 7 (2007): 659–72.

Chapter 5. Bodily Defense

Abbott, N. J. "Blood–Brain Barrier Structure and Function and the Challenges for CNS Drug Delivery." *Journal of Inherited Metabolic Disease* 36 (2013): 437–49.

Pardridge, W. M. "Drug Delivery to the Brain." *Journal of Cerebral Blood Flow and Metabolism* 17 (1997): 713–31.

Patel, M. M., B. R. Goyal, S. V. Bhadada, J. S. Bhatt, and A. F. Amin. "Getting into the Brain." *CNS Drugs* 23 (2009): 35–58.

Chapter 7. Traveling Particles

Blackwell, B. R., K. J. Wooten, M. D. Buser, B. J. Johnson, G. P. Cobb, and P. N. Smith. "Occurrence and Characterization of Steroid Growth Promoters Associated with Particulate Matter Originating from Beef Cattle Feedyards." *Environmental Science and Technology* 49 (2015): 8796–803.

Hamra, G. B., N. Guha, A. Cohen, F. Laden, O. Raaschou-Nielsen, J. M. Samet, P. Vineis, et al. "Outdoor Particulate Matter Exposure and

Lung Cancer: A Systematic Review and Meta-Analysis." *Environmental Health Perspectives* 122 (2014): 906–11. doi:10.1289/ehp.1408092.

Jonker, M. T. O., and L. van Mourik. "Exceptionally Strong Sorption of Infochemicals to Activated Carbon Reduces Their Bioavailability to Fish." *Environmental Toxicology and Chemistry* 33 (2014): 493–99.

Liu, L., B. Urch, R. Poon, M. Szyszkowicz, M. Speck, D. R. Gold, A. J. Wheeler, et al. "Effects of Ambient Coarse, Fine, and Ultrafine Particles and Their Biological Constituents on Systemic Biomarkers: A Controlled Human Exposure Study." *Environmental Health Perspectives* 123 (2015): 534–40. doi:10.1289/ehp.1408387.

McEachran, A. D., B. R. Blackwell, J. D. Hanson, K. J. Wooten, G. D. Mayer, S. B. Cox, and P. N. Smith. "Antibiotics, Bacteria, and Antibiotic Resistance Genes: Aerial Transport from Cattle Feed Yards via Particulate Matter." *Environmental Health Perspectives* 123 (2015): 337–43. doi:10.1289/ehp.1408555.

Sandstrom, T., and B. Forsberg. "Desert Dust: An Unrecognized Source of Dangerous Air Pollution?" *Epidemiology* 19 (2008): 808–9.

Valavanidis, A., K. Fiotakis, and T. Vlachogianni. "Airborne Particulate Matter and Human Health: Toxicological Assessment and Importance of Size and Composition of Particles for Oxidative Carcinogenic Mechanisms." *Journal of Environmental Science and Health Part C: Environmental Carcinogenesis Ecotoxicology Reviews* 26 (2008): 339–62. doi:10.1080/10590500802494538.

Vignati, D. A., T. Dworak, B. Ferrari, B. Koukal, J. L. Loizeau, M. Minouflet, M. I. Camusso, S. Polesello, and J. Dominik. "Assessment of the Geochemical Role of Colloids and Their Impact on Contaminant Toxicity in Freshwaters: An Example from the Lambro-Po System (Italy)." *Environmental Science and Technology* 39 (2005): 489–97.

Von Essen, S. G., and B. W. Auvermann. "Health Effects from Breathing Air Near CAFOs for Feeder Cattle or Hogs." *Journal of Agromedicine* 10 (2005): 55–64.

Chapter 8. Toxins, Poisons, and Venoms

Fry, B. G. "From Genome to 'Venome': Molecular Origin and Evolution of the Snake Venom Proteome Inferred from Phylogenetic Analysis of Toxin Sequences and Related Body Proteins." *Genome Research* 15 (2005): 403–20.

Chapter 9. Metals: Gift and Curse

Clarkson, T. W. "The Three Modern Faces of Mercury." *Environmental Health Perspectives* 110 (2002): 11–23.

Diamond, G. L., P. E. Goodrum, S. P. Felter, and W. L. Ruoff. "Gastrointestinal Absorption of Metals." *Drug and Chemical Toxicology* 21 (1998): 223–51.

Hakim, J. "The Story of the Atom." *American Federation of Teachers* (2002): 12–25.

Le, T. T., W. J. Peijnenburg, A. J. Hendriks, and M. G. Vijver. "Predicting Effects of Cations on Copper Toxicity to Lettuce (*Lactuca sativa*) by the Biotic Ligand Model." *Environmental Toxicology and Chemistry* 31 (2012): 355–59.

Lessler, M. A. "Lead and Lead Poisoning from Antiquity to Modern Times." *Ohio Journal of Science* 88 (1988): 78–84.

Nicole, W. "Evolutionary Selection for Arsenic Resistance: The Case of the Atacamenos of the Andes Highlands." *Environmental Health Perspectives* 121 (2013): A31.

Niyogi, S., and C. H. Wood. "Biotic Ligand Model, a Flexible Tool for Developing Site-Specific Water Quality Guidelines for Metals." *Environmental Science and Technology* 38 (2004): 6177–92.

Paquin, P. R., J. W. Gorsuch, S. Apte, G. E. Batley, K. C. Bowles, P. G. C. Campbell, C. G. Delos, et al. "The Biotic Ligand Model: A Historical Overview." *Comparative Biochemistry and Physiology Part C* 133 (2002): 3–35.

Przygoda, G., J. Feldmann, and W. Cullen. "The Arsenic Eaters of Styria: A Different Picture of People Who Were Chronically Exposed to Arsenic." *Applied Organometallic Chemistry* 15 (2001): 457–62.

Chapter 10. Combustion

Ashby, J., and D. Paton. "The Influence of Chemical Structure on the Extent and Sites of Carcinogenesis for 522 Rodent Carcinogens and 55 Different Human Carcinogen Exposures." *Mutation Research* 286 (1993): 3–74.

Creton, S., M. J. Aardema, P. L. Carmichael, J. S. Harvey, F. L. Martin, R. F. Newbold, M. R. O'Donovan, et al. "Cell Transformation Assays for Prediction of Carcinogenic Potential: State of the Science and Future Research Needs." *Mutagenesis* 27 (2012): 93–101.

DeVita, V. T. Jr., and E. Chu. "A History of Cancer Chemotherapy." *Cancer Research* 68 (2008): 8643–53.

Melicow, M. M. "Percivall Pott (1712–1788): 200th Anniversary of First Report of Occupation-Induced Cancer of Scrotum in Chimney Sweepers (1775)." *Urology* 6 (1975): 745–49.

Miller, E. C. "Some Current Perspectives on Chemical Carcinogenesis in Humans and Experimental Animals: Presidential Address." *Cancer Research* 38 (1978): 1479–96.

Miller, J. A. "Carcinogenesis by Chemicals: An Overview—G. H. A. Clowes Memorial Lecture." *Cancer Research* 30 (1970): 559–76.

Triolo, V. A., and I. L. Riegel. "The American Association for Cancer Research, 1907–1940. Historical Review." *Cancer Research* 21 (1961): 137–67.

Chapter 11. Drugs and the Toxicology of Addiction

Blum, K., J. G. Cull, E. R. Braverman, and D. E. Comings. "Reward Deficiency Syndrome." *American Scientist* 84 (1996): 132–45.

Blum, K., A. L. C. Chen, J. Giordano, J. Borsten, T. J. H. Chen, M. Hauser, T. Simpatico, J. Femino, E. R. Braverman, and D. Barh. "The Addictive Brain: All Roads Lead to Dopamine." *Journal of Psychoactive Drugs* 44 (2012): 134–43.

De Pasquale, A. "Pharmacognosy: The Oldest Modern Science." *Journal of Ethnopharmacology* 11 (1984): 1–16.

Glander, K. E. "Nonhuman Primate Self-Medication with Wild Plants." Chap. 12 in *Eating on the Wild Side: The Pharmacologic, Ecologic, and Social Implications of Using Noncultigens*. Tucson, AZ: University of Arizona Press, 1994.

Hardy, K., S. Buckley, M. J. Collins, A. Estalrrich, D. Brothwell, L. Copeland, A. García-Tabernero, et al. "Neanderthal Medics? Evidence for Food, Cooking, and Medicinal Plants Entrapped in Dental Calculus." *Naturwissenschaften* 99 (2012): 617–26. doi:10.1007/s00114-012-0942-0. Epub 2012 Jul 18.

Huffman, M. A. "Current Evidence for Self-Medication in Primates: A Multidisciplinary Perspective." *Yearbook of the Journal of Physical Anthropology* 40 (1997): 171–200.

Jones, A. W. "Early Drug Discovery and the Rise of Pharmaceutical Chemistry." *Drug Testing and Analysis* 3 (2011): 337–44.

Nichols, D. E. "Hallucinogens." *Pharmacology and Therapeutics* 101 (2004): 131–81.

Quaglio, G., A. Fornasiero, P. Mezzelani, S. Morechini, F. Lugoboni, and A. Lechi. "Anabolic Steroids: Dependence and Complications of Chronic Use." *Internal and Emergency Medicine* 4 (2009): 289–96.

Sullivan, R. J., and E. H. Hagen. "Psychotropic Substance-Seeking: Evolutionary Pathology or Adaptation?" *Addiction* 97 (2002): 389–400.

Tan, R. S., and M. C. Scally. "Anabolic Steroid-Induced Hypogonadism: Towards a Unified Hypothesis of Anabolic Steroid Action." *Medical Hypotheses* 72 (2009): 723–28.

Westermeyer, J. "The Pursuit of Intoxication: Our 100-Century-Old

Romance with Psychoactive Substances." *American Journal of Drug and Alcohol Abuse* 14 (1988): 175–87.

Wood, R. I. "Reinforcing Aspects of Androgens." *Physiology and Behavior* 83 (2004): 279–89.

Chapter 12. 70,000 Years of Pesticides

Bouwman, H., H. van den Berg, and H. Kylin. "DDT and Malaria Prevention: Addressing the Paradox." *Environmental Health Perspectives* 119 (2011): 744–47.

Dunlap, T. R. "Science as a Guide in Regulating Technology: The Case of DDT in the United States." *Social Studies of Science* 8 (1978): 265–85.

McWilliams, J. E. "'The Horizon Opened Up Very Greatly': Leland O. Howard and the Transition to Chemical Insecticides in the United States, 1894–1927." *Agricultural History* 82 (2008): 468–95.

O'Shaughnessy, P. T. (2008) "Parachuting Cats and Crushed Eggs: The Controversy over the Use of DDT to Control Malaria." *American Journal of Public Health* 98: 1940–48.

Smith, A. E., and D. M. Secoy. "Forerunners of Pesticides in Classical Greece and Rome." *Journal of Agricultural and Food Chemistry* 23 (1975): 1050–55.

Chapter 13. Origins of Regulation

Botting, J. "The History of Thalidomide." *Drug News & Perspectives* 15 (2002): 604–11.

Gaughan, A. "Harvey Wiley, Theodore Roosevelt, and the Federal Regulation of Food and Drugs." Student paper, Harvard Law School, 2004. http://nrs.harvard.edu/urn-3:HUL.InstRepos:8852144.

Herbst, A. L., D. C. Poskanzer, S. J. Robboy, L. Friedlander, and R. E. Scully. "Prenatal Exposure to Stilbestrol: A Prospective Comparison of Exposed Female Offspring with Unexposed Controls." *New England Journal of Medicine* 292 (1975): 334–39.

Herbst, A. L., H. Ulfelder, and D. C. Poskanzer. "Adenocarcinoma of the Vagina." *New England Journal of Medicine* 284 (1971): 878–81.

Janssen, W. F. "The Story of the Laws behind the Labels." *FDA Consumer.* US Food and Drug Administration, 1981. http://www.fda.gov/AboutFDA/WhatWeDo/History/Overviews/ucm056044.htm.

Klimko, K. "FDA's Contradictory Decisions Related to the Delaney Clause." Student paper, Harvard Law School, 2011. http://nrs.harvard.edu/urn-3:HUL.InstRepos:8963872.

Neltner, T. G., N. R. Kulkarni, H. M. Alger, M. V. Maffini, E. D. Bongard,

N. D. Fortin, and E. D. Olson. "Navigating the U.S. Food Additive Regulatory Program." *Comprehensive Reviews in Food Science and Food Safety* 10 (2011): 342–68.

Nicole, W. "Secret Ingredients: Who Knows What's in Your Food?" *Environmental Health Perspectives* 121 (2013): A126–33.

Quinn, R. "Rethinking Antibiotic Research and Development: World War II and the Penicillin Collaborative." *American Journal of Public Health* 103 (2013): 426–34.

US Food and Drug Administration. "Guidance for Industry: Frequently Asked Questions about GRAS." US Department of Health and Human Services, 2004. http://www.fda.gov/Food/GuidanceRegulation/GuidanceDocumentsRegulatoryInformation/IngredientsAdditives GRASPackaging/ucm061846.htm.

Wax, P. M. "Elixirs, Diluents, and the Passage of the 1938 Federal Food, Drug, and Cosmetic Act." *Annals of Internal Medicine* 122 (1995): 456–61.

Chapter 14. Legislating for Health

Asch, P. "Food Safety Regulation: Is the Delaney Clause the Problem or Symptom?" *Policy Sciences* 23 (1990): 97–110.

Baker, M. "Insights from the Structure of Estrogen Receptor into the Evolution of Estrogens: Implications for Endocrine Disruption." *Biochemical Pharmacology* 82 (2011): 1–8.

Blair, R. M., H. Fang, W. S. Branham, B. S. Hass, S. L. Dial, C. L. Moland, W. Tong, L. Shi, R. Perkins, and D. M. Sheehan. "The Estrogen Receptor Binding Affinities of 188 Natural and Xenochemicals: Structural Diversity of Ligands." *Toxicological Sciences* 54 (2000): 138–53.

Creton, S., M. J. Aardema, P. L. Carmichael, J. S. Harvey, F. L. Martin, R. F. Newbold, M. R. O'Donovan, et al. "Cell Transformation Assays for Prediction of Carcinogenic Potential: State of the Science and Future Research Needs." *Mutagenesis* 27, no. 1 (Jan 2012): 93–101. doi:10.1093/mutage/ger053.

Fagan, D. "The Learning Curve." *Nature* 490 (25 October 2012): 462–65.

Junod, S. W. "Sugar: A Cautionary Tale." US Food and Drug Administration, 2003. http://www.fda.gov/aboutfda/whatwedo/history/product regulation/selectionsfromfdliupdateseriesonfdahistory/ucm091680 .htm, accessed January 1, 2016.

Levenson, A. S., and V. C. Jordan. "The First Hormone-Responsive Breast Cancer Cell Line." *Cancer Research* 57 (1997): 3071–78.

Melnick, R., G. Lucier, M. Wolfe, R. Hall, G. Stancel, G. Prins, M. Gallo, et al. "Summary of the National Toxicology Program's Report of the

Endocrine Disruptors Low-Dose Peer Review." *Environmental Health Perspectives* 110 (2002): 427–31.
Vandenberg, L. N., T. Colborn, T. B. Hayes, J. J. Heindel, D. R. Jacobs Jr., D.-H. Lee, T. Shioda, et al. "Hormones and Endocrine-Disrupting Chemicals: Low-Dose Effects and Nonmonotonic Dose Responses." *Endocrine Reviews* 33 (2012): 1–78.
Welshons, W. V., K. A. Thayer, B. M. Judy, J. A. Taylor, E. M. Curran, and F. S. vom Saal. "Large Effects from Small Exposures: I. Mechanisms for Endocrine-Disrupting Chemicals with Estrogenic Activity." *Environmental Health Perspectives* 111 (2003): 994–1006.
Yager, J. D., and N. E. Davidson. "Estrogen Carcinogenesis in Breast Cancer." *New England Journal of Medicine* 354 (2006): 270–82.

Chapter 15. POPS and *Silent Spring*

Adeola, F. O. "Boon or Bane? The Environmental and Health Impacts of Persistent Organic Pollutants (POPs)." *Human Ecology Review* 11 (2004): 27–35.
American Chemical Society. "Rachel Carson's *Silent Spring*: National Historic Chemical Landmarks." American Chemical Society, 2013. http://www.acs.org/content/acs/en/education/whatischemistry/landmarks/rachel-carson-silent-spring.html.
Ballschmiter, K., R. Hackenberg, W. M. Jarman, and R. Looser. "Man-Made Chemicals Found in Remote Areas of the World: The Experimental Definition for POPs." *Environmental Science & Pollution Research* 9 (2002): 274–88.
Bard, S. M. "Global Transport of Anthropogenic Contaminants and the Consequences for the Arctic Marine Ecosystem." *Marine Pollution Bulletin* 38 (1999): 359–79.
Bonefeld-Jorgensen, E. C. "Biomonitoring in Greenland: Human Biomarkers of Exposure and Effects—A Short Review." *Rural and Remote Health* 10 (2010): 1362. http://www.rrh.org.au/articles/showarticlenew.asp?ArticleID=1362.
Bowerman, W. W., D. A. Best, T. G. Grubb, G. M. Zimmerman, and J. P. Giesy. "Trends of Contaminants and Effects in Bald Eagles of the Great Lakes Basin." *Environmental Monitoring and Assessment* 53 (1998): 197–212.
Donaldson, S. G., J. Van Oostdam, C. Tikhonov, M. Feeley, B. Armstrong, P. Ayotte, O. Boucher, et al. "Environmental Contaminants and Human Health in the Canadian Arctic." *Science of the Total Environment* 408 (2010): 5165–234. doi:10.1016/j.scitotenv.2010.04.059. Epub 2010 Aug 21.

Chapter 16. Toxic Toiletries

Boxall, A. B., M. A. Rudd, B. W. Brooks, D. J. Caldwell, K. Choi, S. Hickmann, E. Innes, et al. "Pharmaceuticals and Personal Care Products in the Environment: What Are the Big Questions?" *Environmental Health Perspectives* 120 (2012): 1221–29. doi:10.1289/ehp.1104477. Epub 2012 May 30.

Brausch, J. M., and G. M. Rand. "A Review of Personal Care Products in the Aquatic Environment: Environmental Concentrations and Toxicity." *Chemosphere* 82 (2011): 1518–32.

Brooks, B. W. "Fish on Prozac (and Zoloft): Ten Years Later." *Aquatic Toxicology* 151 (2014): 61–67.

Brooks, B. W., C. K. Chambliss, J. K. Stanley, A. Ramirez, K. E. Banks, R. D. Johnson, and R. J. Lewis. "Determination of Select Antidepressants in Fish from an Effluent-Dominated Stream." *Environmental Toxicology and Chemistry* 24 (2005): 464–69.

Brooks, B. W., C. M. Foran, S. M. Richards, J. Weston, P. K. Turner, J. K. Stanley, K. R. Solomon, M. Slattery, and T. W. La Point. "Aquatic Ecotoxicology of Fluoxetine." *Toxicology Letters* 142 (2003): 169–83.

Brooks, B. W., P. K. Turner, J. K. Stanley, J. J. Weston, E. A. Glidewell, C. M. Foran, M. Slattery, T. W. La Point, and D. B. Huggett. "Waterborne and Sediment Toxicity of Fluoxetine to Select Organisms." *Chemosphere* 52 (2003): 135–42.

Fabbri, E. "Pharmaceuticals in the Environment: Expected and Unexpected Effects on Aquatic Fauna." *Annals of the New York Academy of Sciences* 1340 (2015): 20–28.

Gonzalez-Rey, M., N. Tapie, K. Le Menach, M.-H. Devier, H. Budzinski, and M. J. Bebianno. "Occurrence of Pharmaceutical Compounds and Pesticides in Aquatic Systems." *Marine Pollution Bulletin* 96 (2015): 384–400.

James-Todd, T., M. B. Terry, J. Rich-Edwards, A. Deierlein, and R. Senie. "Childhood Hair Product Use and Earlier Age at Menarche in a Racially Diverse Study Population: A Pilot Study." *Annals of Epidemiology* 21 (2011): 461–65. doi:10.1016/j.annepidem.2011.01.009. Epub 2011 Mar 21.

Kessler, R. "More than Cosmetic Changes: Taking Stock of Personal Care Product Safety." *Environmental Health Perspectives* 123 (2015): A120–27. doi:10.1289/ehp.123-A120.

Larsson, J. D. G., C. de Pedro, and N. Paxeus. "Effluent from Drug Manufactures Contain Extremely High Levels of Pharmaceuticals." *Journal of Hazardous Materials* 148 (2007): 751–55. Epub 2007 Jul 6.

Larsson, J. D. G. "Pollution from Drug Manufacturing: Review and Perspectives." *Philosophical Transactions of the Royal Society B* 369 (2014). doi:10.1098/rstb.2013.0571.

Lubick, N. "India's Drug Problem." *Nature* 457 (2009): 640–41. doi:10.1038/457640a.

Matthiessen, P., and L. Weltje. "A Review of the Effects of Azole Compounds in Fish and Their Possible Involvement in Masculinization of Wild Fish Populations." *Critical Reviews in Toxicology* 45 (2015): 453–67. doi:10.3109/10408444.2015.1018409. Epub 2015 Apr 21.

Myers, S. L., C. Z. Yang, G. D. Bittner, K. L. Witt, R. R. Tice, and D. D. Baird. "Estrogenic and Anti-Estrogenic Activity of Off-the-Shelf Hair and Skin Care Products." *Journal of Exposure Science and Environmental Epidemiology* 25 (2015): 271–77. doi:10.1038/jes.2014.32. Epub 2014 May 21.

Oaks, J. L., M. Gilbert, M. Z. Virani, R. T. Watson, C. U. Meteyer, B. A. Rideout, H. L. Shivaprasad, et al. "Diclofenac Residues as the Cause of Vulture Population Decline in Pakistan." *Nature* 427 (12 February 2004): 630–33.

Overturf, M. D., J. C. Anderson, Z. Pandelides, L. Beyger, and D. A. Holdway. "Pharmaceuticals and Personal Care Products: A Critical Review of the Impacts on Fish Reproduction." *Critical Reviews in Toxicology* 45 (2015): 469–91. doi:10.3109/10408444.2015.1038499. Epub 2015 May 6.

Sumpter, J. P., R. L. Donnachie, and A. C. Johnson. "The Apparently Very Variable Potency of the Anti-Depressant Fluoxetine." *Aquatic Toxicology* 151 (2014): 57–60.

Ternes, T. A., A. Joss, and H. Siegrist. "Scrutinizing Pharmaceuticals and Personal Care Products in Wastewater Treatment." *Environmental Science & Technology* 38 (2004): 392A–399A.

Tiwary, C. M. "Premature Sexual Development in Children Following the Use of Estrogen- or Placenta-Containing Hair Products." *Clinical Pediatrics* 37 (1998): 733–39.

Wise, L. A., J. R. Palmer, D. Reich, Y. C. Cozier, and L. Rosenberg. "Hair Relaxer Use and Risk of Uterine Leiomyomata in African-American Women." *American Journal of Epidemiology* 175 (2012): 432–40. doi:10.1093/aje/kwr351.

Chapter 17. Determining Sex: Chemicals and Reproduction

Barber, L. B., A. M. Vajda, C. Douville, D. O. Norris, and J. H. Writer. "Fish Endocrine Disruption Responses to a Major Wastewater Treatment Facility Upgrade." *Environmental Science & Technology* 46 (2012): 2121–31.

Buck Louis, G. M., M. A. Clooney, and C. M. Peterson. "The Ovarian Dysgenesis Syndrome." *Journal of Developmental Origins of Health and Disease* 2 (2011): 25–35.

Devlin, R. H., and Y. Nagahama. "Sex Determination and Sex Differentiation in Fish: An Overview of Genetic, Physiological, and Environmental Influences." *Aquaculture* 208 (2002): 191–364.

Georges, A., T. Ezaz, A. E. Quinn, and S. D. Sarre. "Are Reptiles Predisposed to Temperature-Dependent Sex Determination?" *Sexual Development* 4 (2010): 7–15.

Grasman, K. A., P. F. Scanlon, and G. A. Fox. "Reproductive and Physiological Effects of Environmental Contaminants in Fish-Eating Birds of the Great Lakes: A Review of Historical Trends." *Environmental Monitoring and Assessment* 53 (1998): 117–45.

Guillette, L. J. Jr., "Endocrine Disrupting Contaminants—Beyond the Dogma." *Environmental Health Perspectives* 114 (2006): 9–12.

Hamlin, H. J., and L. J. Guillette Jr. "Birth Defects in Wildlife: The Role of Environmental Contaminants as Inducers of Reproductive and Developmental Dysfunction." *Systems Biology in Reproductive Medicine* 56 (2010): 113–21.

Harris, C. A., P. B. Hamilton, T. J. Runnalis, V. Vinciotti, A. Henshaw, D. Hodgson, T. S. Coe, S. Jobling, C. R. Tyler, and J. P. Sumpter. "The Consequences of Feminization in Breeding Groups of Wild Fish." *Environmental Health Perspectives* 119 (2011): 306–11.

Hayes, T. B., V. Khoury, A. Narayan, M. Nazir, A. Park, T. Brown, L. Adame, et al. "Atrazine Induces Complete Feminization and Chemical Castration in Male African Clawed Frogs (*Xenopus laevis*)." *Proceedings of the National Academy of Sciences USA* 107 (2010): 4612–17.

Jiménez, R., F. J. Barrionuevo, and M. Burgos. "Natural Exceptions to Normal Gonad Development in Mammals." *Sexual Development* 7 (2013): 147–62.

Kidd, K., P. J. Blanchfield, K. H. Mills, V. P. Palace, R. E. Evans, J. M. Lazorchak, and R. W. Flick. "Collapse of a Fish Population after Exposure to a Synthetic Estrogen." *Proceedings of the National Academy of Sciences USA* 104 (2007): 8897–901.

LeBlanc, G. A. "Are Environmental Sentinels Signaling?" *Environmental Health Perspectives* 103 (1995): 888–90.

MacLachlan, N. J. "Ovarian Disorders in Domestic Animals." *Environmental Health Perspectives* 73 (1987): 27–33.

Marley, C. L., H. McCalman, S. Buckingham, D. Downes, and M. T. Abberton. "A Review of the Effect of Legumes on Ewe and Cow Fertility." *IBERS Legumes and Fertility Review* (2011): 1–30.

McLachlan, J. A. "Environmental Signaling: What Embryos and Evolution Teach Us about Endocrine Disrupting Chemicals." *Endocrine Reviews* 22 (2001): 319–41.

Nakamura, M. "Sex Determination in Amphibians." *Seminars in Cell and Developmental Biology* 20 (2009): 271–82.

Nakamura, M. "Morphological and Physiological Studies on Gonadal Sex Differentiation in Teleost Fish." *Aqua-BioScience Monographs* 6 (2013): 1–47.

Newbold, R. R., and J. A. McLachlan. "Vaginal Adenosis and Adenocarcinoma in Mice Exposed Prenatally or Neonatally to Diethylstilbestrol." *Cancer Research* 42 (1982): 2003–11.

Ryan, B. C., and J. G. Vandenbergh. "Intrauterine Position Effects." *Neuroscience & Biobehavioral Reviews* 26 (2002): 665–78.

Smith, C. A., K. N. Roeszler, T. Ohnesorg, D. M. Cummins, P. G. Farlie, T. J. Doran, and A. H. Sinclair. "The Avian Z-Linked Gene DMRT1 Is Required for Male Sex Determination in the Chicken." *Nature* 461 (10 September 2009): 267–71.

Swain, A., and R. Lovell-Badge. "Mammalian Sex Determination: A Molecular Drama." *Genes & Development* 13 (1999): 755–67.

Vaiman, D., and E. Pailhoux. "Mammalian Sex Reversal and Intersexuality: Deciphering the Sex-Determination Cascade." *Trends in Genetics* 16 (2000): 488–94.

Zheng, Z., and M. J. Cohn. "Developmental Basis of Sexually Dimorphic Digit Ratios." *Proceedings of the National Academy of Sciences USA* 108 (2011): 16289–94.

Chapter 18. The Earliest Exposure: Transgenerational Toxicology

Anway, M. D., A. S. Cupp, M. Uzumcu, and M. K. Skinner. "Epigenetic Transgenerational Actions of Endocrine Disruptors and Male Fertility." *Science* 308 (2005): 1466–69.

Anway, M. D., C. Leathers, and M. K. Skinner. "Endocrine Disruptor Vinclozolin-Induced Epigenetic Transgenerational Adult-Onset Disease." *Endocrinology* 147 (2006): 5515–23. Epub 2006 Sep 14.

Bannister, A. J., and T. Kouzarides. "Regulation of Chromatin by Histone Modifications." *Cell Research* 21 (2011): 381–95.

Barker, D. K. P., C. Osmond, P. D. Winter, B. Margetts, and S. J. Simmonds. "Weight in Infancy and Death from Ischaemic Heart Disease." *Lancet* 334 (9 September 1989): 577–80.

Carey, N. "Beyond DNA: Epigenetics." In *The Epigenetic Revolution: How Modern Biology Is Rewriting Our Understanding of Genetics, Disease, and Inheritance*. New York: Columbia University Press, 2012. Excerpted in *Natural History* Online (n.d.). www.naturalhistorymag.com/features/142195/beyond-dna-epigenetics.

Esteller, M. "Non-Coding RNAs in Human Disease." *Nature Reviews Genetics* 12 (2011): 861–74.

Heijmans, B. T., E. W. Tobi, A. D. Stein, H. Putter, G. J. Blauw, E. S. Susser, P. E. Slagboom, and L. H. Lumey. "Persistent Epigenetic Differences Associated with Prenatal Exposure to Famine in Humans." *Proceedings of the National Academy of Sciences USA* 105 (2008): 17046–9. doi:10.1073/pnas.0806560105. Epub 2008 Oct 27.

Jirtle, R. L., and M. K. Skinner. "Environmental Epigenomics and Disease Susceptibility." *Nature Reviews Genetics* 8 (2007): 253–62.

Louis, G. M., M. A. Cooney, and C. M. Peterson. "The Ovarian Dysgenesis Syndrome." *Journal of Developmental Origins of Health and Disease* 2 (2011): 25–35.

Lumey, L. H., M. B. Terry, L. Delgado-Cruzata, Y. Liao, Q. Wang, E. Susser, I. McKeague, and R. M. Santella. "Adult Global DNA Methylation in Relation to Pre-Natal Nutrition." *International Journal of Epidemiology* 41 (2012): 116–23. doi:10.1093/ije/dyr137. Epub 2011 Sep 29.

Martin, G. M. "Epigenetic Drift in Aging Identical Twins." *Proceedings of the National Academy of Sciences USA* 102 (2005): 10413–14.

Nilsson, E. E., and M. K. Skinner. "Environmentally Induced Epigenetic Transgenerational Inheritance of Disease Susceptibility." *Translational Research* 165 (2015): 12–17.

Painter, R. C., T. J. Roseboom, and O. P. Bleker. "Prenatal Exposure to the Dutch Famine and Disease in Later Life: An Overview." *Reproductive Toxicology* 20 (2005): 345–52.

Poulsen, P., M. Esteller, A. Vaag, and M. F. Fraga. "The Epigenetic Basis of Twin Discordance in Age-Related Diseases." *Pediatric Research* 61 (2007): 38R–42R.

Prins, G. S. "Estrogen Imprinting: When Your Epigenetic Memories Come Back to Haunt You." *Endocrinology* 149 (2008): 5919–21.

Reik, W., W. Dean, and J. Walter. "Epigenetic Reprogramming in Mammalian Development." *Science* 293 (10 August 2001): 1089–92.

Reik, W., and J. Walter. "Genomic Imprinting: Parental Influence on the Genome." *Nature Reviews Genetics* 2 (2001): 21–32.

Schmidt, C. W. "Uncertain Inheritance Transgenerational Effects of Environmental Exposures." *Environmental Health Perspectives* 121 (2013): A298–A303.

Uzumcu, M., A. M. Zama, and E. Oruc. "Epigenetic Mechanisms in the Actions of Endocrine-Disrupting Chemicals: Gonadal Effects and Role in Female Reproduction." *Reproduction in Domestic Animals* 47 (2012): 338–47. doi:10.1111/j.1439-0531.2012.02096.x.

Xin, F., M. Susiarjo, and M. S. Bartolomei. "Multigenerational and Transgenerational Effects of Endocrine Disrupting Chemicals: A Role for Altered Epigenetic Regulation?" *Seminars in Cell Developmental Biology* 43 (2015): 66–75. doi:10.1016/j.semcdb.2015.05.008 [Epub ahead of print].

Zhang, X., and S. M. Ho. "Epigenetics Meets Endocrinology." *Journal of Molecular Endocrinology* 46 (2011): R11–R32.

Chapter 19. Natural Toxins Revisited

Aquzzi, A., and J. Falsig, eds. "Prion Propagation, Toxicity, and Degradation." *Nature Neuroscience* 15 (2012): 936–39.

Espelund, M., and D. Klaveness. "Botulism Outbreaks in Natural Environments—An Update." *Frontiers in Microbiology* 5 (11 June 2014): 287. doi:10.3389/fmicb.2014.00287.

Friend, M., and J. C. Franson, eds. "Avian Botulism." Chap. 38 in *Field Manual of Wildlife Diseases: General Field Procedures and Diseases of Birds.* USGS Technology Report 1999–001. Washington, DC: USGS Biological Resources Division, 1999. http://www.nwhc.usgs.gov/publications/field_manual/field_manual_of_wildlife_diseases.pdf.

Saunders, S. E., S. L. Bartelt-Hunt, and J. C. Bartz. "Occurrence, Transmission, and Zoonotic Potential of Chronic Wasting Disease." *Emerging Infectious Diseases* 18 (2012): 369–76. doi:10.3201/eid1803.110685.

Saunders, S. E., S. L. Bartelt-Hunt, and J. C. Bartz. "Prions in the Environment: Occurrence, Fate, and Mitigation." *Prion* 2 (2008): 162–69. Epub 2008 Oct 26.

Saunders, S. E., J. C. Bartz, and S. L. Bartelt-Hunt. "Soil-Mediated Prion Transmission: Is Local Soil-Type a Key Determinant of Prion Disease Incidence?" *Chemosphere* 87 (2012): 661–67.

Texas Department of Insurance. "Poison Ivy, Oak, and Sumac Factsheet." HS04-064B(3-07). Texas Department of Insurance, Division of Workers' Compensation (TDI/DWC) [n.d.].

Chapter 20. Chemical Resistance

Arias, C. A., and B. E. Murray. "Antibiotic-Resistant Bugs in the 21st Century—A Clinical Super-Challenge." *New England Journal of Medicine* 360 (2009): 439–43. doi:10.1056/NEJMp0804651.

Bass, C., and L. M. Field. "Gene Amplification and Insecticide Resistance." *Pest Management Science* 67 (2011): 886–90.

Bouki, C., D. Venieri, and E. Diamadopoulos. "Detection and Fate of Antibiotic Resistant Bacteria in Wastewater Treatment Plants: A Review." *Ecotoxicology and Environmental Safety* 91 (2013): 1–9. doi:10.1016/j.ecoenv.2013.01.016. Epub 2013 Feb 13.

Brown, M. G., and D. L. Balkwill. "Antibiotic Resistance in Bacteria Isolated from the Deep Terrestrial Subsurface." *Microbial Ecology* 57 (2009): 484–93.

Davies, J., and D. Davies. "Origins and Evolution of Antibiotic Resistance." *Microbiology and Molecular Biology Reviews* 74 (2010): 417–33. doi:10.1128/MMBR.00016-10.

Ffrench-Constant, R. H. "The Molecular Genetics of Insecticide Resistance." *Genetics* 194 (2013): 807–15. doi:10.1534/genetics.112.141895.

Gonzalez-Rey, M., N. Tapie, K. Le Menach, M. H. Dévier, H. Budzinski, and M. J. Bebianno. "Occurrence of Pharmaceutical Compounds and Pesticides in Aquatic Systems." *Marine Pollution Bulletin* 96 (2015): 384–400. doi:10.1016/j.marpolbul.2015.04.029. Epub 2015 May 18.

Hartley, C. J., R. D. Newcomb, R. J. Russell, C. G. Yong, J. R. Stevens, D. K. Yeates, J. La Salle, and J. G. Oakeshott. "Amplification of DNA from Preserved Specimens Shows Blowflies Were Preadapted for the Rapid Evolution of Insecticide Resistance." *Proceedings of the National Academy of Sciences* 103 (2006): 8757–62.

Heuer, H., and K. Smalla. "Horizontal Gene Transfer between Bacteria." *Environmental Biosafety Research* 6 (2007): 3–13. Epub 2007 Oct 26.

Kim, S., and D. S. Aga. "Potential Ecological and Human Health Impacts of Antibiotics and Antibiotic-Resistant Bacteria from Wastewater Treatment Plants." *Journal of Toxicology and Environmental Health, Part B Critical Reviews* 10 (2007): 559–73.

Levine, D. P. "Vancomycin: A History." *Clinical Infectious Diseases* 42 (Supplement 1) (2006): S5–S12. doi:10.1086/491709.

Levy, S. B., and B. Marshall. "Antibacterial Resistance Worldwide: Causes, Challenges, and Responses." *Nature Medicine* 10 (2004): S122–29.

Li, X., M. A. Schuler, and M. R. Berenbaum. "Molecular Mechanisms of Metabolic Resistance to Synthetic and Natural Xenobiotics." *Annual Review of Entomology* 52 (2007): 231–53. doi:10.1146/annurev.ento.51.110104.151104.

Martinez, J. L. "Environmental Pollution by Antibiotics and by Antibiotic Resistance Determinants." *Environmental Pollution* 157 (2009): 2893–902. doi:10.1016/j.envpol.2009.05.051. Epub 2009 Jun 27.

Mell, J. C., and R. J. Redfield. "Natural Competence and the Evolution of

DNA Uptake Specificity." *Journal of Bacteriology* 196 (2014): 1471–83.

Meyers, J. I., M. Gray, and B. D. Foy. "Mosquitocidal Properties of IgG Targeting the Glutamate-Gated Chloride Channel in Three Mosquito Disease Vectors (Diptera: Culicidae). *Journal of Experimental Biology* 218 (2015): 1487–95.

Nielsen, K. M., P. J. Johnsen, D. Bensasson, and D. Daffonchio. "Release and Persistence of Extracellular DNA in the Environment." *Environmental Biosafety Research* 6 (2007): 37–53. Epub 2007 Sep 12.

Olofsson, S. K., P. Geli, D. I. Andersson, and O. Cars. "Pharmacodynamic Model to Describe the Concentration-Dependent Selection of Cefotaxime-Resistant *Escherichia coli*." *Antimicrobial Agents and Chemotherapy* 49 (2005): 5081–91.

Palecchi, L., A. Bartolini, F. Paradisi, and G. M. Rosolini. "Antibiotic Resistance in the Absence of Antimicrobial Use: Mechanisms and Implications." *Expert Review of Anti-Infective Therapy* 6 (2008): 725–32. doi:10.1586/14787210.6.5.725.

Palumbi, S. R. "Humans as the World's Greatest Evolutionary Force." *Science* 293 (2001): 1786–90.

Quinn, R. "Rethinking Antibiotic Research and Development: World War II and the Penicillin Collaborative." *American Journal of Public Health* 103 (2013): 426–34. doi:10.2105/AJPH.2012.300693. Epub 2012 Jun 14.

Roberts, J. A., P. Kruger, D. L. Paterson, and J. Lipman. "Antibiotic Resistance—What's Dosing Got to Do with It?" *Critical Care Medicine* 36 (2008): 2433–40. doi:10.1097/CCM.0b013e318180fe62.

Sexton, S. E., Z. Lei, and D. Zilberman. "The Economics of Pesticides and Pest Control." *International Review of Environmental and Resource Economics* 1 (2007): 271–326.

Tilman, D., J. Fargione, B. Wolff, C. D'Antonio, A. Dobson, R. Howarth, D. Schindler, W. H. Schlesinger, D. Simberloff, and D. Swackhamer. "Forecasting Agriculturally Driven Global Environmental Change." *Science* 292 (2001): 281–84.

Turnidge, J., and D. L. Paterson. "Setting and Revising Antibacterial Susceptibility Breakpoints." *Clinical Microbiology Reviews* 20 (2007): 391–408.

Verraes, C., S. Van Boxstael, E. Van Meervenne, E. Van Coillie, P. Butaye, B. Catry, M. A. de Schaetzen, et al. "Antimicrobial Resistance in the Food Chain: A Review." *International Journal of Environmental Research and Public Health* 10 (2013): 2643–69.

World Health Organization. Fact sheet No. 270. "Botulism." WHO Media

Centre, August 2013. http://www.who.int/mediacentre/factsheets/fs270/en/, retrieved September 4, 2015.

Zhang, Y., C. Zhang, D. B. Parker, D. D. Snow, Z. Zhou, and X. Li. "Occurrence of Antimicrobials and Antimicrobial Resistance Genes in Beef Cattle Storage Ponds and Swine Treatment Lagoons." *Science of the Total Environment* 463–64 (2013): 631–38. doi:10.1016/j.scitotenv.2013.06.016. Epub 2013 Jul 7.

INDEX

Island Press | Board of Directors